Practice for the End-of-Grade Test

Grade 4

Orlando Austin Chicago New York Toronto London San Diego

Visit *The Learning Site!*
www.harcourtschool.com

Copyright © by Harcourt, Inc.

All rights reserved. No part of this publication may be reproduced or transmitted in any form or by any means, electronic or mechanical, including photocopy, recording, or any information storage and retrieval system, without permission in writing from the publisher.

Permission is hereby granted to individual teachers using the corresponding student's textbook or kit as the major vehicle for regular classroom instruction to photocopy complete pages from this publication in classroom quantities for instructional use and not for resale. Requests for information on other matters regarding duplication of this work should be addressed to School Permissions and Copyrights, Harcourt, Inc., 6277 Sea Harbor Drive, Orlando, Florida 32887-6777. Fax: 407-345-2418.

HARCOURT and the Harcourt Logo are trademarks of Harcourt, Inc., registered in the United States of America and/or other jurisdictions.

Printed in the United States of America

ISBN 0-15-339044-1

4 5 6 7 8 9 10   054   12 11 10 09 08 07 06

Contents

▶ **Number & Operations**
Number & Operations Vocabulary Practice 1
Number & Operations Practice 3
Number & Operations Practice Test 41

▶ **Measurement**
Measurement Vocabulary Practice 49
Measurement Practice 51
Measurement Practice Test 59

▶ **Geometry**
Geometry Vocabulary Practice 63
Geometry Practice 65
Geometry Practice Test 73

▶ **Data Analysis & Probability**
Data Analysis & Probability Vocabulary Practice 77
Data Analysis & Probability Practice 79
Data Analysis & Probability Practice Test 93

▶ **Algebra**
Algebra Vocabulary Practice 99
Algebra Practice 101
Algebra Practice Test 112

▶ **EOG Practice Test** 117

Name _____

NUMBER & OPERATIONS

Vocabulary

Change the underlined part to make the statement true. Use a term from the box.

estimate
factor
inverse operation
multiple
product

1. Mel found the <u>sum</u> of two factors by multiplying them together.

   _____

2. John found a(n) <u>exact answer</u> for 32 times 18 by rounding before he multiplied.

   _____

3. Ann knew that 15 is a(n) <u>factor</u> of 3 because 3 times 5 equals 15.

   _____

4. Elaine checked her division work by using the <u>Commutative Property</u>, which is multiplication.

   _____

Match the terms to the correct definitions. Record the numbers in the magic square. To check your work, make sure the sums for the columns, rows, and diagonals are all the same.

A	B	C
D	E	F
G	H	I

A. remainder
B. dividend
C. divisor
D. quotient
E. benchmark number
F. divisible
G. estimate
H. divide
I. compatible numbers

1. to find how many items will be in each group
2. the number that divides the dividend
3. the answer, not including the remainder, that results from dividing
4. the amount left over when a number cannot be divided evenly
5. a known number of things that helps you understand the size or amount of a different number of things
6. numbers that are easy to compute with
7. able to be divided so that the quotient is a whole number and the remainder is zero
8. to find an answer that is close to the exact amount
9. the number that is to be divided

Number & Operations Vocabulary Practice   Practice for the EOG Test   1

Name _____

NUMBER & OPERATIONS

Vocabulary

An analogy shows a relationship between two terms that are related in a similar way.

Examples: <u>Dog</u> is to <u>puppy</u> as <u>cat</u> is to <u>kitten</u>.

<u>Plane</u> is to <u>air</u> as <u>boat</u> is to <u>water</u>.

Complete each analogy.

1. Addition is to subtraction as _____ is to division.

2. A number of parts is to the total number of parts as _____ is to denominator.

3. Simplest form is to equivalent fraction as _____ is to improper fraction.

Use the clues to complete the number puzzle.

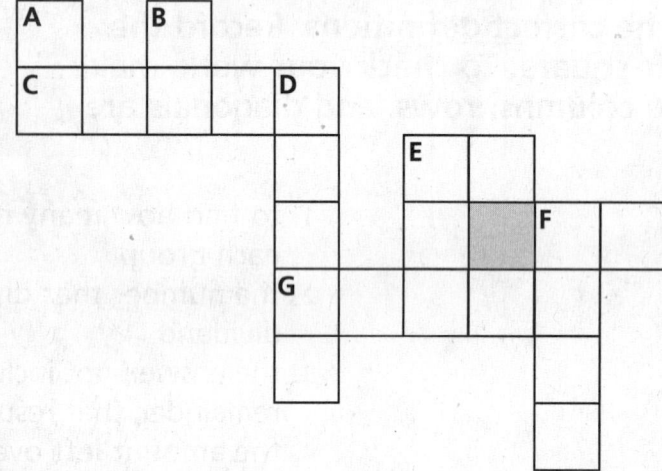

ACROSS
C. Write the greater number: 64,305 or 64,298.
E. Which number is divisible by 3: 13, 24, or 53?
F. Which is a multiple of 2: 16, 19, or 23?
G. Write the value of the digit 4 in 45,829.

DOWN
A. Write the product of six times six.
B. Write the remainder of 83 ÷ 14.
D. Round 56,136 to the nearest ten.
E. What compatible number would you use to compute 4 × 248?
F. Estimate the sum of 841 and 162 by using rounding.

2 Practice for the EOG Test

Number & Operations Vocabulary Practice

Name _____

NUMBER & OPERATIONS

1. The land area of North Carolina is forty-eight thousand, seven hundred eighteen square miles. Write the number of square miles in standard form.

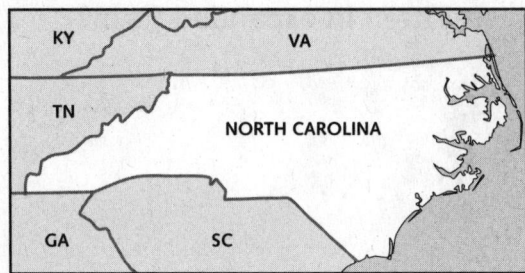

_____

2. Along North Carolina's Outer Banks, Cape Lookout National Seashore includes about 28,500 acres of undeveloped barrier islands. Write the number of acres in word form.

_____

3. On a video game, Kristin's score was shown like this.

Ten Thousands	Thousands	Hundreds	Tens	Ones
●●●●●●●	●●●●		●●●●●●●	●●●

Write her score in standard form.

_____

4. **Explain It** Dave drew a dot in each box of this 10-by-10 grid.

How many 10-by-10 grids of dots will he need to show 10,000 dots? Explain how you know.

_____

© Harcourt

1.01 Develop number sense for rational numbers 0.01 through 99,999. a) Connect model, number word, and number using a variety of representations.
Source for North Carolina State Standards: North Carolina Department of Public Instruction

Number & Operations Practice   Practice for the EOG Test 3

Name _____

NUMBER & OPERATIONS

5. The land areas of states that border North Carolina are shown in the table.

Areas of States Bordering North Carolina

State	Area (in square miles)
Georgia	57,919
South Carolina	30,111
Tennessee	41,219
Virginia	39,598

Which state's land area has a digit 9 with a value of 9,000? has a digit 2 with a value of 200?

_____

6. Carla's score on a ring-toss game is one thousand, two hundred forty-five. If you add a ten thousands digit that is 2 times the hundreds digit to Carla's score, you will get Beth's score. What is Beth's score written in expanded form?

_____

7. The table shows the heights of the highest points in five states.

Highest Point

State	Height (in feet)
Georgia	4,784
North Carolina	6,684
South Carolina	3,560
Tennessee	6,643
Virginia	5,729

Which state's highest point rounds to 6,000 feet?

TIP To which place do you want to round? Is the digit to its right *less than 5*, or is it *5 or more*?

8. **Explain It** Brad is playing a number game. He drew these number cards.

| 2 | 6 | 4 | 9 | 0 |

If Brad uses each digit only once, what is the *least* number that is greater than 10,000 that he can make? Explain how you know.

_____
_____
_____
_____
_____

1.01 Develop number sense for rational numbers 0.01 through 99,999. b) Build understanding of place value (hundredths through ten thousands).

4 Practice for the EOG Test Number & Operations Practice

Name _____

NUMBER & OPERATIONS

9. The table shows the heights of some mountains in the United States.

Heights of United States Mountains

Mountain	State	Height (in feet)
Mt. Elbert	Colorado	14,433
Mt. Rainier	Washington	14,410
Mt. Whitney	California	14,494
Mt. Williamson	California	14,375

Which mountain is the tallest?

_____

10. The table shows the average depths of some oceans.

Depths of Oceans

Ocean	Average Depth (in feet)
Atlantic	11,730
Arctic	3,407
Indian	12,598
Pacific	12,925

Write the names of the oceans in order from *greatest* average depth to *least* average depth.

_____

11. Mr. and Mrs. Bryson are planning to buy a used car. They listed the colors, the miles driven, and the prices of four cars.

Used Cars

Color	Miles Driven	Price
Red	36,489	$13,500
Green	37,800	$12,800
Black	38,030	$9,999
Blue	37,225	$10,450

List the colors of the cars from the *least* number of miles driven to the *greatest* number of miles driven.

_____

12. **Explain It** Jen and Ernesto are playing a number-guessing game. Jen said that she is thinking of a number that is greater than the *greatest* three-digit number and less than the *least* five-digit number. Describe all the numbers Jen could be thinking of.

_____
_____
_____
_____

1.01 Develop number sense for rational numbers 0.01 through 99,999. c) Compare and order rational numbers.

Number & Operations Practice Practice for the EOG Test 5

Name _____

NUMBER & OPERATIONS

13. Alan and Linda save pennies. There are 20 pennies in Alan's jar.

Are there *more* than or *less* than 200 pennies in Linda's jar?

_____

14. Mrs. Chin needs paper clips for the math handout. There are 100 paper clips in the small box. *About* how many paper clips are in the large box?

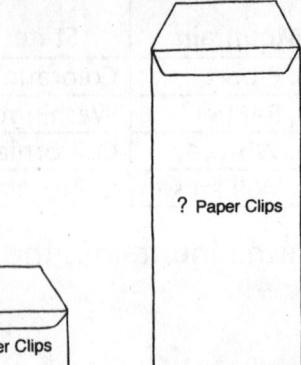

Are there about 50, about 500, or about 5,000 paper clips in the large box?

_____

15. What would be a good benchmark to estimate the number of books in your school library? Explain.

TIP What is a known number of books you can use to find a greater number of books?

16. **Explain It** Amy counted 20 flowers in one section of her flower garden. *About* how many flowers are in the garden in all? Explain how you know.

1.01 Develop number sense for rational numbers 0.01 through 99,999. d) Make estimates of rational numbers in appropriate situations.

6 **Practice for the EOG Test** **Number & Operations Practice**

Name _____

NUMBER & OPERATIONS

17. Neil ate 3 muffins. What fraction of the muffins did he eat?

_____

18. Rebecca drew this design. Shade $\frac{1}{4}$ of her design. Then write a fraction for the unshaded part.

_____

19. Kristen used $\frac{5}{6}$ pound of apples, $\frac{3}{4}$ pound of grapes, and $\frac{1}{2}$ pound of peaches to make a fruit salad. Write two true statements to compare the fraction of a pound of each fruit she used. Use <, >, or = in each statement.

TIP How does making a model help you compare fractions with unlike denominators?

_____

20. Explain It Mike says that because 8 is greater than 6, $\frac{5}{8}$ is greater than $\frac{5}{6}$. Do you agree or disagree with him? Draw a diagram to explain your answer.

_____
_____
_____
_____

Maintains (3) 1.05 Use area or region models and set models of fractions to explore part-whole relationships.
a) Represent fractions concretely and symbolically (halves, fourths, thirds, sixths, eighths). b) Compare and order fractions (halves, fourths, thirds, sixths, eighths) using models and benchmark numbers (zero, one-half, one); describe comparisons.

Number & Operations Practice   Practice for the EOG Test

Name _____

NUMBER & OPERATIONS

21. The first model shows $\frac{1}{2}$ shaded. Shade to show equivalent fractions in the other models. Write a fraction for the part you shaded.

$\frac{1}{2}$ _____ _____

22. Heather drew a number line to model some numbers between 0 and 3. What mixed number should she write at point A?

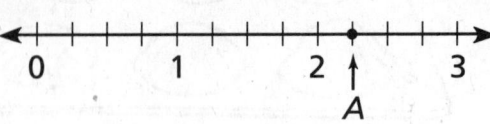

_____

23. Write a mixed number and a fraction for the shaded part of the picture.

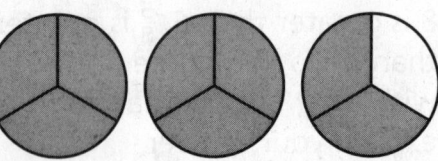

_____   _____
mixed number   **fraction**

How can you change the picture to show 3 wholes? Show your work with pictures or words.

_____

24. Explain It The children ate 3 whole pizzas and $\frac{1}{2}$ of another pizza for lunch. Draw a picture to show the pizzas they ate. Write a mixed number and a fraction for the picture.

_____   _____
mixed number   **fraction**

How many halves does your diagram show?

_____

Maintains (3) 1.05 Use area or region models and set models of fractions to explore part-whole relationships. c) Model and describe common equivalents, especially relationships among halves, fourths, and eighths, and thirds and sixths. d) Understand that the fractional relationships that occur between zero and one also occur between every two consecutive whole numbers. e) Understand and use mixed numbers and their equivalent fraction forms.

8 Practice for the EOG Test   Number & Operations Practice

Name _____

NUMBER & OPERATIONS

25. Katie used a hundredths model to show the total number of inches of rain reported in the last seven days.

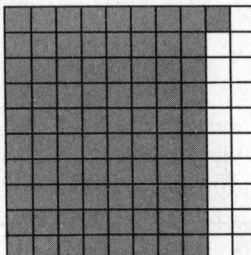

How many inches of rain does her model show?

_____

26. Horatio bought 6.75 pounds of apples to make apple tarts for the school fair.

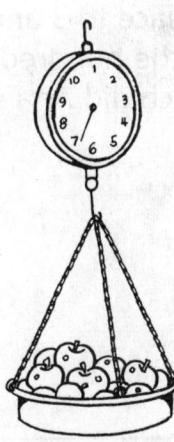

Write the number of pounds of apples he bought in word form.

_____

27. Blake's house is one and twenty-five hundredths miles from school. Write the number of miles from Blake's house to his school in standard form.

TIP How can you use a hundredths model to write this number?

_____

28. Explain It Draw a model to show that 1.5 and 1.50 are equivalent. Explain your model.

_____
_____
_____
_____

1.01 Develop number sense for rational numbers 0.01 through 99,999. a) Connect model, number word, and number using a variety of representations.

Number & Operations Practice

Practice for the EOG Test 9

Name _____

NUMBER & OPERATIONS

29. Jake spent $0.42 at the grocery store. To find how much Ana spent, change the digit in the tenths place to 3 and change the digit in the hundredths place to 8. How much did Ana spend?

_____

30. Kaila drew a number line to show some numbers between 0 and 0.5. What decimal and fraction should she write at the arrows?

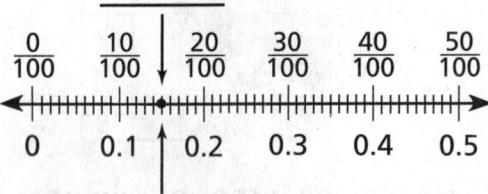

_____

31. Leah drew five cards with numbers and one card with a decimal point. If Leah uses each card only once, what is the *greatest* number with a digit in the hundredths place that she can write?

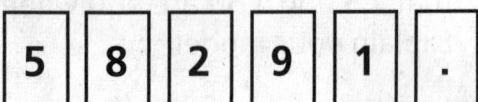

_____

32. **Explain It** Pam did a science project on plant growth. She recorded the weekly growth for her plant in the table below.

Pam's Plant

Week	Growth (in centimeters)
1	0.25
2	0.30
3	0.23

She said her plant grew three tenths of an inch in Week 2. Is she correct? Explain how you know.

_____
_____
_____
_____

1.01 Develop number sense for rational numbers 0.01 through 99,999. b) Build understanding of place value (hundredths through ten thousands).

10 Practice for the EOG Test   Number & Operations Practice

Name _____

NUMBER & OPERATIONS

33. Andrew and his friends are planning to go hiking. The table shows the lengths and the difficulty levels of the trails they can hike.

Hiking Trails

Trail	Length (in miles)	Difficulty Level
Spruce	2.58	Easy
Pine	2.63	Difficult
Fir	2.50	Easy

They decide to hike the longest easy trail. Which trail will they hike?

_____

34. In the long-jump competition, Eddie finished in first place, Bob finished in second place, and Craig finished in third place. Eddie jumped 3.25 meters. Craig jumped 3.14 meters. Write a distance that Bob could have jumped.

TIP How can you use a place-value chart or a number line to help you solve this problem?

_____

35. Maria and her family recorded the number of miles they traveled each day on their vacation. On Saturday, they traveled 145.56 miles; on Sunday, 129.23 miles; on Monday, 72.79 miles; and on Tuesday, 145.83 miles. List the days from *least* number of miles traveled to *greatest* number of miles traveled.

36. **Explain It** Mrs. Laurel wrote this statement on the board.

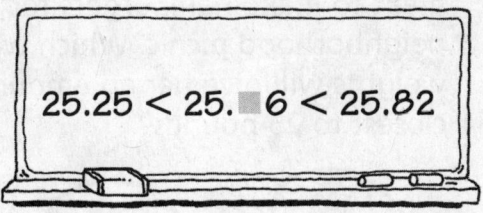

$25.25 < 25.\blacksquare 6 < 25.82$

List all the possible digits for the missing digit. Explain your choices.

_____

1.01 Develop number sense for rational numbers 0.01 through 99,999. c) Compare and order rational numbers.

Number & Operations Practice   Practice for the EOG Test   11

Name _____

NUMBER & OPERATIONS

37. Rosa and her family drove 48.5 miles to Chimney Rock Park from Brevard. To the nearest mile, how far did they drive?

38. Pat measured the length of the bulletin board.

1.9 meters

He wants to put a ribbon border around the outside edge of the bulletin board. Will he need *more* than or *less* than 8 meters of ribbon?

39. Jessica needs 25 pounds of ground meat to make hamburgers for a neighborhood picnic. Which two weights will give her an amount closest to 25 pounds?

14.61 pounds   13.28 pounds

9.92 pounds

40. **Explain It** Madison says that 0.65 is between 0.5 and 1.0 but closer to 0.5. Do you agree or disagree? Draw a diagram to explain your answer.

1.01 Develop number sense for rational numbers 0.01 through 99,999. d) Make estimates of rational numbers in appropriate situations.

12 Practice for the EOG Test   Number & Operations Practice

Name _____

NUMBER & OPERATIONS

41. Four children are playing a number game. They take turns tossing six balls. Each time a ball sticks to a section of the game board, the child scores the number of points shown in that section. Mark's six tosses are shown in the diagram.

TIP How does what you know about place value help you solve this problem?

a. What was Mark's score? _____

b. Nathan's score was twenty-one thousand, thirteen. Write his score in standard form.

_____

c. To find Opal's score, look at Nathan's score. Change the digit in the thousands place to a digit that is 3 times the digit in the tens place. Then change the digit in the tens place to show 40 more. What was Opal's score?

_____

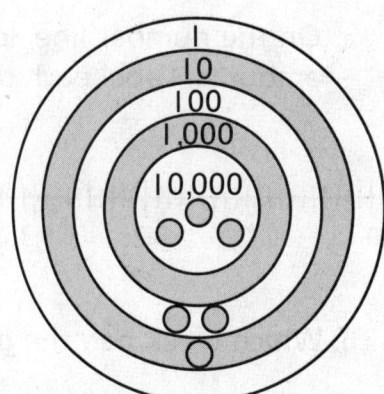

d. Pat tossed six balls. Three balls landed in one section, and three balls landed in another section. Draw a diagram to show one way the balls could have landed. Write the score for the model you draw.

Score _____

Explain how you know that the score and the model you drew match.

_____

1.01 Develop number sense for rational numbers 0.01 through 99,999. a) Connect model, number word, and number using a variety of representations. b) Build understanding of place value (hundredths through ten thousands).

Number & Operations Practice **Practice for the EOG Test 13**

Name _____

NUMBER & OPERATIONS

42. Charlotte is keeping a record of weekly rainfall totals for a science experiment. The table shows how many inches of rain have fallen in the last 5 weeks.

Week	1	2	3	4	5
Rainfall (in inches)	0.57	0.04	0.25	0.40	0.72

a. On the number line, locate the decimals for the rainfall amounts. Label each decimal with the number of the week.

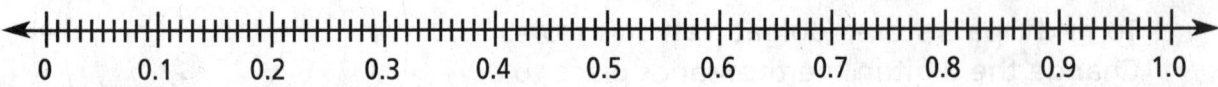

b. Which week had the *greatest* amount of rainfall?
_____

c. In which week was the amount of rainfall closest to 0.5 inch?
_____

d. Order the decimals for the rainfall amounts from *least* to *greatest*.
_____

e. The amount of rainfall in Week 6 was *greater* than the amount of rainfall in Week 4 but *less* than the amount of rainfall in Week 1. Write one possible rainfall amount for Week 6. Explain how you know your answer is correct.

_____

_____

_____

1.01 Develop number sense for rational numbers 0.01 through 99,999. c) Compare and order rational numbers.

14 Practice for the EOG Test

Number & Operations Practice

Name _____

NUMBER & OPERATIONS

1. Janice is designing a garden. She is deciding how to plant 12 bushes. Write two multiplication equations that describe her two designs. Then write two related division equations.

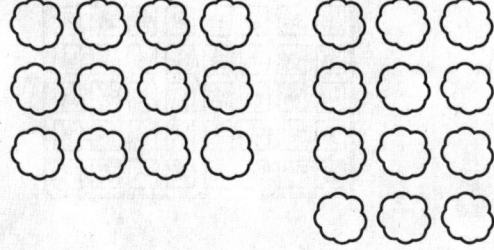

_____

_____

2. Simon has 48 colored markers. If he shares them with 5 friends equally, how many colored markers will he share with each friend? How many will be left over?

3. Pamela is making muffins. She is using pans that make 3 rows of 4 muffins each. If she makes 4 pans of muffins, how many muffins will she make?

TIP How can the Associative Property help you find the product?

4. **Explain It** Luis works at a grocery store arranging boxes. He put some boxes in 2 rows of 10, and then in 4 rows of 5, but both ways were too wide for the aisle. Using the same number of boxes, draw pictures to show two other ways he could arrange the boxes to take up less space.

Maintains (3) 1.03 Develop fluency with multiplication from 1 × 1 to 12 × 12 and division up to two-digit by one-digit numbers using: a) Strategies for multiplying and dividing numbers. c) Relationships between operations.

Number & Operations Practice Practice for the EOG Test

Name _____

NUMBER & OPERATIONS

1. Carla from the Sunrise Bakery is ordering eggs to make baked goods. If she buys 30 cartons with 12 eggs in each carton, how many eggs will she have?

_____

2. Workers at Pratt's Nursery are planting seeds in flats. Each flat holds 48 plants. If they plant 52 flats, how many plants will they have?

_____

3. Al's Bike Shop has 32 bikes on display. The shop also has 25 packages of neon stickers for sale. Each package contains 15 stickers and costs $3.00. How many stickers are there in all?

TIP What information do you need to solve the problem?

4. **Explain It** Tyesha wants to order 23 boxes of math books. There are 36 books in each box. Complete the order form for Tyesha.

Greenfield Book Store Order Form

Name of Item	Number of Boxes	Number of Books per Box	Total Number of Books
Math			

Explain whether you used paper and pencil, mental computation, or a calculator to find the total number of books she plans to order.

_____
_____
_____
_____
_____

1.02 Develop fluency with multiplication and division: a) Two-digit by two-digit multiplication (larger numbers with calculator).

16 Practice for the EOG Test   Number & Operations Practice

Name _____

NUMBER & OPERATIONS

5. The fourth graders are setting up 560 chairs for parents' night. They are setting up the chairs in 14 sections. How many chairs will be in each section?

6. The school library is moving to a new building. Mr. Jones must pack 809 books. Boxes for the books hold 18 books. How many boxes will he need?

7. Laura's spelling scores for five tests were 85, 92, 78, 84, and 91. What is the mean score for these five tests?

8. **Explain It** Washington Elementary School ordered 660 new coat hooks for the coat racks. The hooks came in 15 boxes. How many coat hooks were in each box?

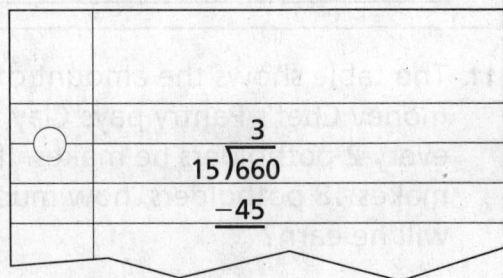

Is this first estimate *too high* or *too low*? Explain how you know. Then divide.

TIP To find the mean, what is the first step you take?

1.02 Develop fluency with multiplication and division: b) Up to three-digit by two-digit division (larger numbers with a calculator).

Number & Operations Practice   Practice for the EOG Test   **17**

Name _____

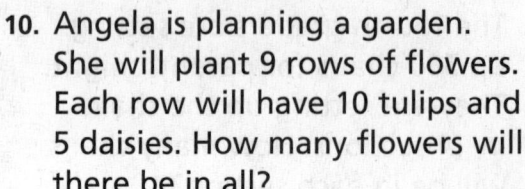

9. A factory makes 70 rubber balls each hour. How many balls can the factory make in one 40-hour work week?

TIP How can you use a basic fact and a pattern to help you solve the problem?

_____

10. Angela is planning a garden. She will plant 9 rows of flowers. Each row will have 10 tulips and 5 daisies. How many flowers will there be in all?

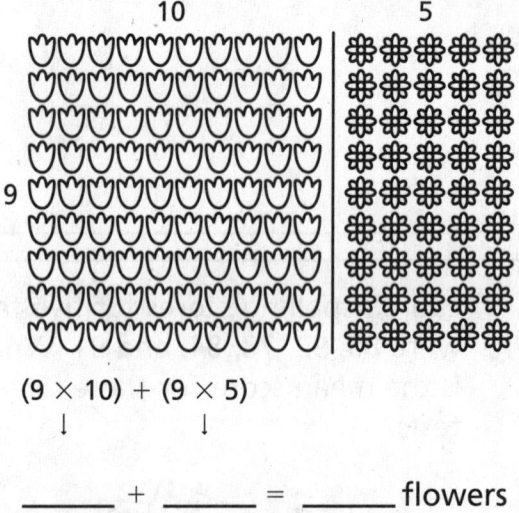

$(9 \times 10) + (9 \times 5)$

_____ + _____ = _____ flowers

11. The table shows the amount of money Chef's Pantry pays Clay for every 2 potholders he makes. If he makes 18 potholders, how much will he earn?

Potholders Clay Makes

Number of Potholders	Money Earned
2	$8.00
4	$16.00
6	$24.00
10	$40.00
12	$48.00

_____

12. **Explain It** During a bird count, birdwatchers sighted 200 Northern cardinals in one zip code area of North Carolina. If birdwatchers from 50 zip code areas reported sighting the same number of cardinals, *about* how many cardinals were sighted?

Number of Northern cardinals in 50 zip code areas:
$200 \times 50 = 1,000$

Describe the error. Write the correct answer.

_____
_____
_____
_____

1.02 Develop fluency with multiplication and division: c) Strategies for multiplying and dividing numbers.

18 Practice for the EOG Test   Number & Operations Practice

Name _____

NUMBER & OPERATIONS

13. Mrs. Vandenboom made 200 meatballs for a party for 50 guests. How many meatballs per guest is that?

_____

14. An outdoor park sold $7,200 worth of tickets one month during the summer. If the ticket price was $8, how many tickets were sold?

_____

15. Alex is using a rule to separate toothpicks, t, into piles, p, for an art project. Study the table and find a rule. Write the rule as an equation. Then find the missing numbers.

Input	t	72	66	60	54	48	42
Output	p	12	11		9		

TIP How can you check your rule?

Rule: _____

_____

16. **Explain It** North Carolina has 78,000 miles of paved roads. The state wants to assign an equal number of miles to maintenance crews so that no miles are left over. How can it figure out what number to divide by? How many crews would it need?

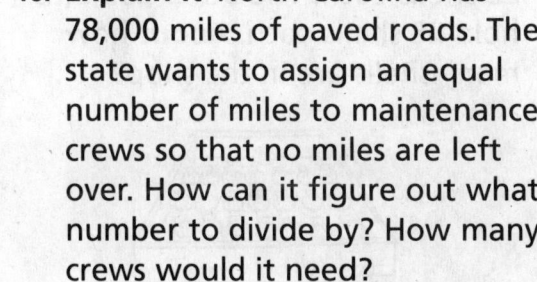

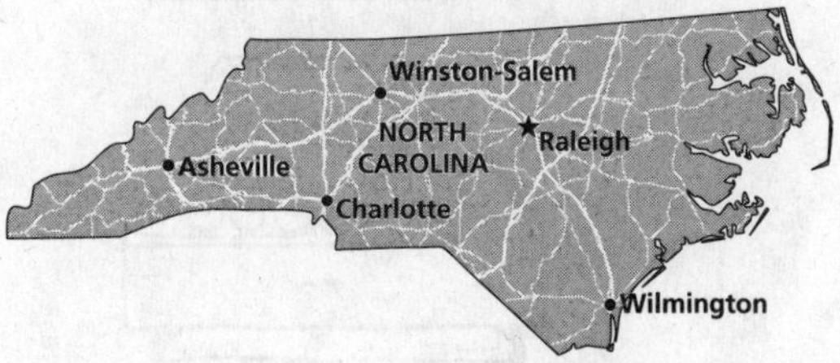

_____

_____

1.02 Develop fluency with multiplication and division: d) Strategies for multiplying and dividing numbers.

Number & Operations Practice Practice for the EOG Test **19**

Name _____

NUMBER & OPERATIONS

17. The North Carolina state tree is the pine tree. If 291 pine trees are growing in 1 acre of a forest, *about* how many pine trees would be growing in a section of forest that is 9 acres?

_____

18. All the fourth-grade classes are going on a field trip. The trip costs $6.75 for each student. If there are 75 students in all, *about* how much money will be collected?

_____

19. The Seaside Surf Shop carries about 508 varieties of shells for sale. About 275 of the bins hold 22 shells each. The other 233 bins hold 28 shells each. *About* how many shells are in the shop?

TIP How can you use rounding to help you estimate?

_____

20. **Explain It** A furniture maker sells a dining room set for $758.89. He estimated he would make $30,000 if he sold 42 sets. Is his estimate reasonable? Explain how you know.

_____
_____
_____
_____
_____
_____

1.02 Develop fluency with multiplication and division: d) Estimation of products and quotients in appropriate situations.

20 Practice for the EOG Test

Number & Operations Practice

Name _____

NUMBER & OPERATIONS

21. Janet has 146 leaves from different trees and shrubs. She puts 3 leaves on each page of a scrapbook. *About* how many pages of her scrapbook does she fill?

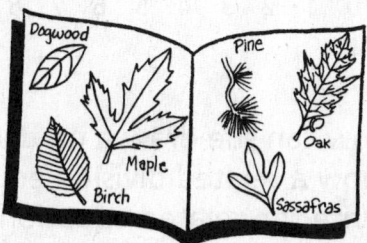

22. The rock observation tower at Wayah Bald in North Carolina is about 260 feet tall. If you climbed 50 feet per minute, *about* how long would it take you to get to the top of the tower?

TIP How can a basic fact and a pattern help you estimate a quotient?

23. Stone steps have been built along the side of a waterfall so hikers can get to the top. There are 778 steps leading up to the observation deck. If there are 2 steps per foot of the height of the waterfall, *about* how tall is the waterfall? Would the exact answer be *more than* or *less than* the estimate?

24. **Explain It** Paul has collected 662 strips from ash trees to mix with other woods to make baskets. If each basket uses 8 strips of ash, how many baskets will have strips of ash in them? Explain how you can use estimation to place the first digit in the quotient. Then solve.

1.02 Develop fluency with multiplication and division: a) Estimation of products and quotients in appropriate situations.

Number & Operations Practice Practice for the EOG Test **21**

Name _____

NUMBER & OPERATIONS

25. Carl collected 12 post cards for each place he visited. If he collected 72 postcards, how many places did he visit?

    TIP How can you use a fact family to solve the problem?

26. Michelle drew this model to show $5 \times 2 = 10$.

    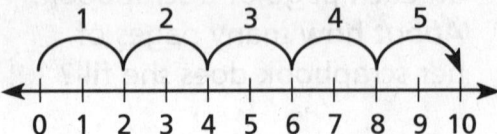

    How can she draw a model to show a related division equation? Write the related equation.

    _____
    _____
    _____

27. Marisol has 36 pictures of classmates to arrange on the bulletin board in groups. How many ways can she arrange them if she has the same number of pictures in each group? How many would be in each group?

28. **Explain It** The Hiking Club logged 261 miles last year. They went on 9 trails. If all the trails were of equal length, what operation can be used to find the length of each trail? What operation can you use to check your answer? Explain.

    _____
    _____
    _____
    _____
    _____

1.02 Develop fluency with multiplication and division: b) Relationships between operations.

22 Practice for the EOG Test   Number & Operations Practice

Name _____

NUMBER & OPERATIONS

29. The city of Maplewood Park is planning new gardens for its city parks. Their landscape designer drew this garden design.

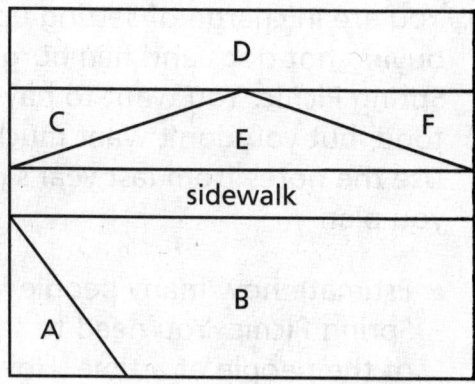

a. In section A, the designer plans to plant 77 coneflowers. What would be a good estimate for the number of coneflowers in section B? Explain.

_____

_____

_____

b. In section D, the designer plans to plant 14 rows of 34 dahlias each. How many dahlias does the designer plan to plant?

_____

c. For sections C, E, and F, the designer has $2,500 to buy plants. He wants to plant 3 different kinds of plants. Make a list of the numbers and kinds of plants he can buy so that the cost of the plants is $2,500 or less. Explain your answer with pictures, words, or numbers.

Plants for Sale

Plants	Price Per Plant
Roses	$28
Peonies	$19
Dahlias	$12
Marigolds	$6
Zinnias	$5

	PS	Comp	Comm
IV			
III			
II			
I			

1.02 Develop fluency with multiplication and division. a) Two-digit by two-digit multiplication (larger numbers with calculator). d) Estimation of products and quotients in appropriate situations.

Number & Operations Practice   Practice for the EOG Test   23

Name _____

NUMBER & OPERATIONS

30. You are in charge of setting up tables and buying hot dogs and hamburgers for the Spring Picnic. You want to have enough food, but you don't want much left over. Use the notes from last year's picnic to help you plan.

    Notes from the Picnic
    - 362 people ate at the picnic.
    - More people ate hamburgers than hot dogs.
    - Hot dogs: 8 per package
    - Hamburgers: 12 per package
    - Tables seat 10 people.

    a. Estimate how many people will eat at the Spring Picnic. You need to seat about half of the people at a time. How many tables should you set up? Explain your answer.

    _____
    _____
    _____
    _____

    b. How many packages of hot dogs and hamburgers should you buy? Explain why you think these are good estimates.

    _____
    _____
    _____
    _____

    c. This year you decide to order packages of cheese slices so that half the hamburger orders can have a slice of cheese. Cheese slices come in packages of 36. How many packages of cheese will you order? Explain your answer.

    _____
    _____
    _____
    _____

	PS	Comp	Comm
IV			
III			
II			
I			

1.02 Develop fluency with multiplication and division. b) Up to three-digit by two-digit division (larger numbers by calculator). d) Estimation of products and quotients in appropriate situations.

24 Practice for the EOG Test

Number & Operations Practice

Name _____

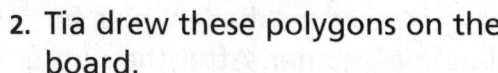

1. Evan cut a batch of brownies into 12 pieces. He ate 2 brownies. Draw a picture and shade it to show the number of brownies that are left. Then write the fraction for the number of brownies that are left.

   _____

2. Tia drew these polygons on the board.

   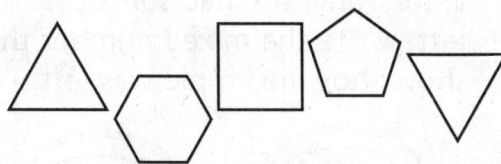

   What fraction of the polygons are not triangles or squares? What are two equivalent fractions for this fraction?

   _____

3. Colin does not have very much brown sugar. He wants to choose the recipe that uses the least amount. One recipe calls for $\frac{1}{4}$ cup and another recipe calls for $\frac{1}{3}$ cup. He chose to make the recipe that uses $\frac{1}{4}$ cup. How did he know, without looking at the measuring cups, that $\frac{1}{4}$ is *less than* $\frac{1}{3}$?

   TIP If one whole is divided into 8 equal parts, are the parts smaller or larger than one whole divided into 4 equal parts?

4. **Explain It** Mrs. Hicks divides the first graders' morning into $\frac{1}{12}$ center time, $\frac{1}{2}$ seat work, $\frac{1}{4}$ story corner work, and $\frac{1}{6}$ board work. Place the fractions on the number line. Is *more* time spent doing story corner work or board work? Explain how you know.

   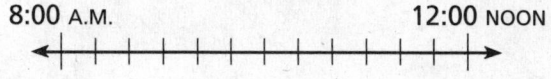

   _____
   _____
   _____
   _____

1.03 Solve problems using models, diagrams, and reasoning about fractions and relationships among fractions involving halves, fourths, eighths, thirds, sixths, twelfths, fifths, tenths, hundredths, and mixed numbers.

Number & Operations Practice   Practice for the EOG Test   25

Name _____

NUMBER & OPERATIONS

5. Ms. Fitzwilliams made 4 pies for a holiday dinner. After the guests were gone, she had some pie left. Write the mixed number that shows how much pie was left.

 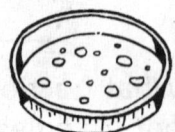

_____

6. The state bird of North Carolina is the cardinal. A male cardinal is about $7\frac{3}{4}$ inches long. Rename the mixed number as a fraction.

_____

7. At recess, George ran $1\frac{1}{10}$ miles and Edna ran $1\frac{7}{10}$ miles. Who ran **more than** $1\frac{3}{5}$ miles?

_____

8. **Explain It** Cafeteria workers put pieces of pie on plates for people to buy. They cut each pie into eighths. At the end of the day, they had 21 pieces of pie left. Did they have closer to 2 or 3 whole pies left? Draw a picture to explain.

_____

1.03 Solve problems using models, diagrams, and reasoning about fractions and relationships among fractions involving halves, fourths, eighths, thirds, sixths, twelfths, fifths, tenths, hundredths, and mixed numbers.

26 Practice for the EOG Test

Number & Operations Practice

Name _____

NUMBER & OPERATIONS

9. Neena invited 5 friends to a pizza party. She decided to order 2 pizzas and divide them each into 6 slices, so that each person could have a total of 2 slices.

   a. Draw the 2 pizzas and shade parts to show the fraction of the pizzas each person could have. Write a fraction for the shaded part.

   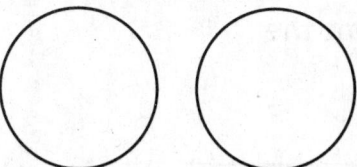

   _____

   b. On Thursday, 2 more friends said they would come. Neena decided to divide the 2 pizzas into 8 slices each. Draw the 2 pizzas and shade parts to show the fraction of the pizzas each person could have. Write a fraction for the shaded part.

   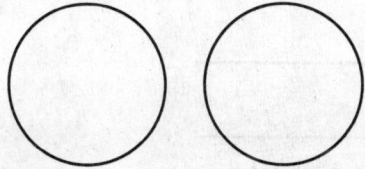

   _____

   c. On Friday, 4 more people said they would come, for a total of 12. Neena thought that she could divide the 2 pizzas into 12 slices each. That way each person could have $\frac{2}{12}$ of the pizzas. Describe her error. Write the correct answer.

   _____

   _____

   d. A large pizza can be cut into 8 slices. If 11 people come, how many pizzas should Neena order so that each person could have the same number of slices? How many slices of pizza could each person have?

   _____

   _____

1.03 Solve problems using models, diagrams, and reasoning about fractions and relationships among fractions involving halves, fourths, eighths, thirds, sixths, twelfths, fifths, tenths, hundredths, and mixed numbers.

Number & Operations Practice   Practice for the EOG Test 27

Name _____

NUMBER & OPERATIONS

10. Amy drew this picture.

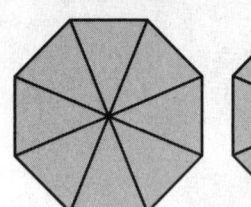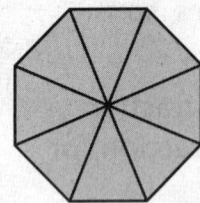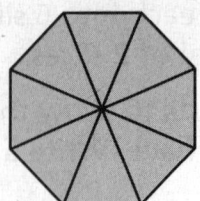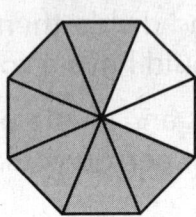

a. Write a fraction and a mixed number to represent the shaded part of Amy's picture.

_____
_____

b. How can you change her picture to show $2\frac{5}{8}$?

_____
_____

c. How can you change her picture to show 4 wholes?

_____
_____

d. How can you change her picture so that it shows a mixed number that is **greater than** $1\frac{1}{4}$ and **less than** $3\frac{3}{8}$? Write the fraction and the mixed number.

_____
_____

e. List the numbers in problems a–d, and order them from **least** to **greatest**.

_____
_____

1.03 Solve problems using models, diagrams, and reasoning about fractions and relationships among fractions involving halves, fourths, eighths, thirds, sixths, twelfths, fifths, tenths, hundredths, and mixed numbers.

28 Practice for the EOG Test   Number & Operations Practice

Name _____

NUMBER & OPERATIONS

1. Corinne and her sister Catherine are saving for a summer vacation. Corinne has $105 so far, and Catherine has $297. How much do they have in all?

_____

2. The Great Smoky Mountains National Park has hills and mountains that range from 875 feet to 6,684 feet. What is the difference between the lowest elevation and the highest elevation?

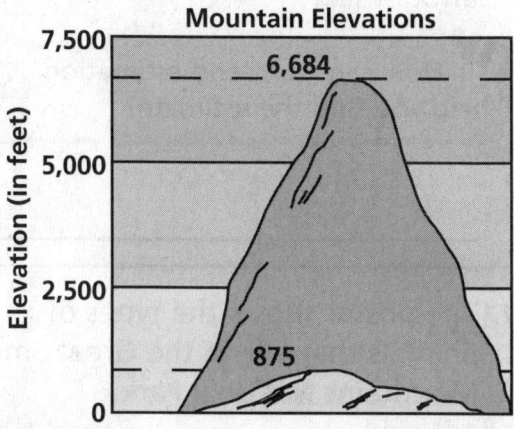

_____

3. Dimitri is buying graham crackers for a camping trip. There are 36 crackers in one box and 75 in another box. Use mental computation to find out how many are in both boxes. How many *more* graham crackers are in one box than in the other?

TIP How can you use the break apart strategy or the make a ten strategy to solve?

_____

4. **Explain It** North Carolina has 350 miles of cold-water stocked trout streams and 3,026 miles of cold-water wild trout streams. How many miles of trout streams does North Carolina have in all? Explain how you can check your answer.

_____
_____
_____

Maintains (3) 1.02 Develop fluency with multi-digit addition and subtraction through 9,999 using: a) Strategies for adding and subtracting numbers. c) Relationships between operations.

Number & Operations Practice   **Practice for the EOG Test** 29

Name _____

NUMBER & OPERATIONS

5. Piedmont and Centennial Elementary Schools are joining together to raise money for a local charity. Piedmont raised $3,243 and Centennial raised $8,523. *About* how much money did they raise altogether?

   TIP How can front-end estimation help you find the estimate?

   _____

6. The Smokies are among the tallest mountains in the Appalachian chain. Mount Le Conte towers to 6,593 feet from its base at 1,292 feet above sea level, making it the tallest mountain in the East. *About* how tall is Mount Le Conte?

   TIP How can rounding help you estimate?

   _____

7. The poster shows the types of animals that live in the Great Smoky Mountains National Park.

   Types of Animals
   - some 66 species of mammals
   - over 230 varieties of birds
   - 50 native fish species
   - more than 80 types of reptiles and amphibians

   About how many more varieties of birds live in the park than types of mammals, reptiles, and amphibians put together?

   _____

8. **Explain It** In 2002, there were 44,610 beagles and 42,571 dachshunds registered with the American Kennel Club. Colin said that 2,039 more beagles were registered than dachshunds. Is his statement reasonable? Explain how you know.

   _____
   _____
   _____
   _____
   _____
   _____
   _____

Maintains (3) 1.02 Develop fluency with multi-digit addition and subtraction through 9,999 using: a) Strategies for adding and subtracting numbers. b) Estimation of sums and differences in appropriate situations.

30 Practice for the EOG Test   Number & Operations Practice

Name _____

NUMBER & OPERATIONS

1. According to Jen's pedometer, she walked $\frac{4}{10}$ of a mile on Monday morning. She walked $\frac{2}{10}$ of a mile on Tuesday morning. How far did she walk in all?

| $\frac{1}{10}$ | $\frac{1}{10}$ | $\frac{1}{10}$ | $\frac{1}{10}$ | | $\frac{1}{10}$ | $\frac{1}{10}$ |

_____

2. A recipe calls for $\frac{1}{8}$ teaspoon of red pepper and $\frac{3}{8}$ teaspoon of white pepper. How much pepper does the recipe call for in all?

_____

3. Lee researched giraffes for his class science project. He made the table below to show what he found.

Giraffe Facts

Parts of a Body	Length (in feet)
Neck	about $6\frac{1}{2}$
Tongue	about $\frac{7}{12}$
Back	about 5
Legs	about 6
Tail	about 3

Lee found that the giraffe is the tallest land animal. Male giraffes may grow to be more than 18 feet tall. *About* how long is the body of a giraffe from the top of the neck to the end of the tail?

TIP What information in the table do you need to answer the problem?

_____

4. **Explain It** Ms. Smith drove $2\frac{1}{4}$ miles to the library and then $1\frac{3}{4}$ miles to the shopping center. Draw a picture to show how far she drove altogether. Explain how you found the answer.

_____
_____
_____
_____
_____
_____

1.04 Develop fluency with addition and subtraction of non-negative rational numbers with like denominators, including decimal fractions through hundredths. a) Develop and analyze strategies for adding and subtracting numbers. b) Estimate sums and differences.

Number & Operations Practice   **Practice for the EOG Test 31**

Name _____

NUMBER & OPERATIONS

5. Sherry measured a board and found it to be $\frac{15}{16}$ foot long. She cut off a piece that measured $\frac{12}{16}$ foot. How long is the piece that was left?

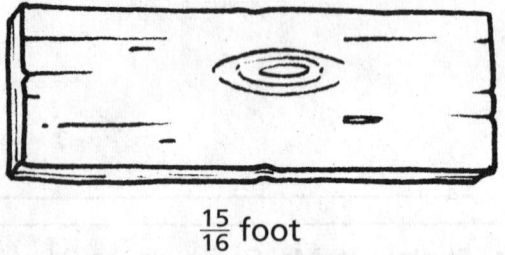

$\frac{15}{16}$ foot

6. David measures out $1\frac{1}{4}$ cups of flour from a canister that contains $7\frac{3}{4}$ cups of flour. How much flour is left in the canister? How do you know that your answer is reasonable? Explain your thinking.

_____
_____
_____
_____

7. Cheryl's recipe for tortilla soup includes these ingredients:

| $1\frac{1}{4}$ cups onion |
| $1\frac{3}{4}$ cups tomato puree |
| $1\frac{1}{4}$ cups sliced chicken breast |
| $\frac{3}{4}$ cup diced avocado |
| $\frac{2}{4}$ cup shredded cheddar cheese |

Her cooking pan holds 8 cups. Is her pan large enough to hold all of the ingredients? Explain.

TIP Which operation should you do first?

_____
_____
_____

8. **Explain It** Tevin said the difference between $3\frac{5}{6}$ and $1\frac{2}{6}$ is $2\frac{1}{2}$. Is he correct? How do you know? Show your work with pictures, words, or numbers.

_____
_____
_____
_____
_____

1.04 Develop fluency with addition and subtraction of non-negative rational numbers with like denominators, including decimal fractions through hundredths. a) Develop and analyze strategies for adding and subtracting numbers. b) Estimate sums and differences. c) Judge the reasonableness of solutions.

32 Practice for the EOG Test

Number & Operations Practice

Name _____

NUMBER & OPERATIONS

9. For a science experiment, Mr. Cooney's fourth-grade class measured the amount of rainfall for 2 days. On the first day, they measured 0.36 inch. On the second day, they measured 0.29 inch. How much did they measure for both days?

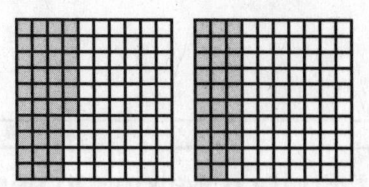

10. Nina saved $25.75 from babysitting. She also saved $38.56 from shoveling snow. How much did she save in all?

TIP What words help you decide what operation to use to solve the problem?

11. Margaret's mother sends her to the store to buy milk, cereal, and bread. The milk costs $3.29. The cereal is on sale for $2.50. The loaf of bread costs $1.99. How much change should Margaret get back if she gives the cashier $20.00?

12. **Explain It** The table shows some of the ten lowest recorded rainfall totals for the month of March from 1924 through 1999. Estimate to find which three amounts will total as close to 1 inch as possible. Then find the total of the three amounts. Explain your reasoning.

March Lowest Rainfall Totals for Wichita, Texas

Year	Amount (in inches)
1925	0.14
1997	0.23
1932	0.29
1972	0.40
1954	0.45

© Harcourt

1.04 Develop fluency with addition and subtraction of non-negative rational numbers with like denominators, including decimal fractions through hundredths. a) Develop and analyze strategies for adding and subtracting numbers. b) Estimate sums and differences.

Number & Operations Practice   **Practice for the EOG Test** 33

Name _____

NUMBER & OPERATIONS

13. Elena rode her bike 3.78 miles. Juan rode for 0.92 mile. *About* how much farther did Elena ride than Juan?

    TIP How can rounding help you answer the question?

    _____

14. The North Carolina state vegetable is the sweet potato. If a bushel of sweet potatoes weighs 32.04 pounds, how much *more* is that than a bushel of green beans weighing 15.61 pounds?

    _____

15. Lisa bought a skirt on sale that cost $15.75 with tax. She also bought a sweatshirt for $22.68 with tax. How much *more* did she pay for the sweatshirt? How do you know that your answer is reasonable? Explain your thinking.

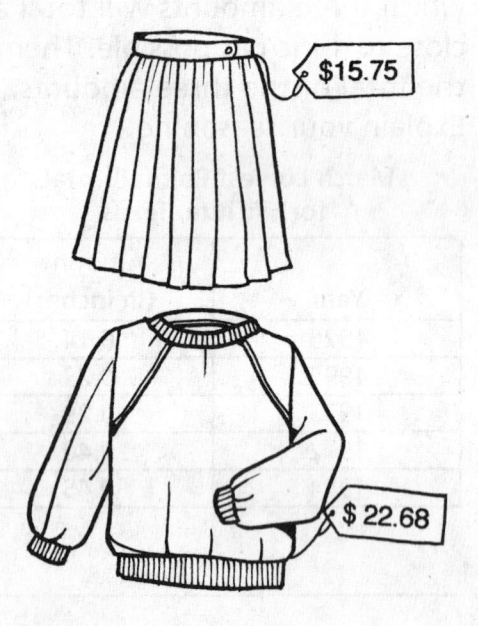

    $15.75

    $22.68

    _____
    _____
    _____

16. **Explain It** Rosita is walking to her grandmother's house, which is 3.5 miles away. Her pedometer says she has walked 1.28 miles so far. She says she has 2.78 miles to go. Describe her error and find the correct answer. Explain using words, pictures, or numbers.

    _____
    _____
    _____

1.04 Develop fluency with addition and subtraction of non-negative rational numbers with like denominators, including decimal fractions through hundredths. a) Develop and analyze strategies for adding and subtracting numbers. b) Estimate sums and differences. c) Judge the reasonableness of solutions.

34 **Practice for the EOG Test** **Number & Operations Practice**

Name _____

NUMBER & OPERATIONS

17. Carol is making a curtain for her kitchen window.

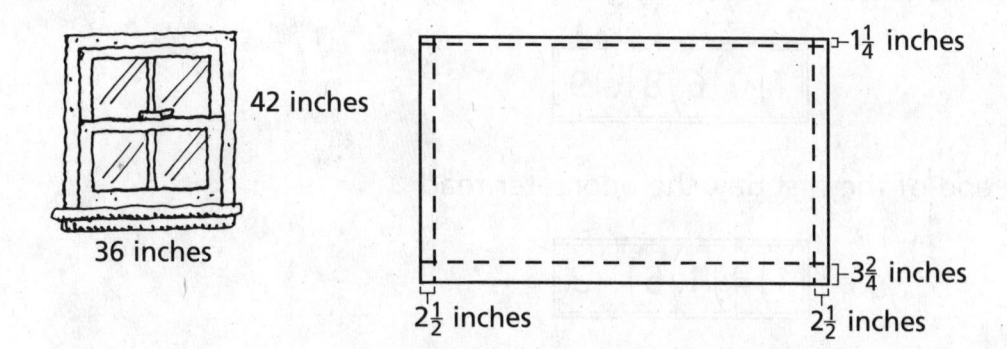

42 inches
36 inches

$1\frac{1}{4}$ inches
$3\frac{2}{4}$ inches
$2\frac{1}{2}$ inches
$2\frac{1}{2}$ inches

- Since the curtain will be gathered, Carol wants the hemmed width to be 2 times the width of the window.
- Carol wants the hemmed length to be 5 inches more than the length of the window.
- Carol plans to sew a $2\frac{1}{2}$-inch hem along each side. Before she cuts the fabric, she needs to add this amount to each side of the hemmed width.
- Carol plans to sew a $1\frac{1}{4}$-inch hem at the top and a $3\frac{2}{4}$-inch hem at the bottom. Before she cuts the fabric, she needs to add these amounts to the hemmed length.

a. What are the width and length of the fabric that she should cut for her curtain? Describe the steps you took to solve the problem. Show your work using words, pictures, or numbers.

b. Carol has two pieces of fabric to choose from. One piece is $62\frac{1}{4}$ inches long, and the other is $51\frac{1}{4}$ inches long. Both pieces are 108 inches wide. Is there a piece long enough for the curtain? If so, which is it? How do you know?

	PS	Comp	Comm
IV			
III			
II			
I			

1.04 Develop fluency with addition and subtraction of non-negative rational numbers with like denominators, including decimal fractions through hundredths. a) Develop and analyze strategies for adding and subtracting numbers.

Number & Operations Practice Practice for the EOG Test **35**

NUMBER & OPERATIONS

18. On a family vacation, Benito kept track of the car's odometer. He took the first reading before they started.

    1 0 6 8 . 9

    a. At the end of the first day, the odometer read

    1 4 1 6 . 3

    How far did they drive on the first day?

    _____

    b. At the end of the second day, the reading was

    2 1 0 7 . 4

    How far did they drive on the second day?

    _____

    c. What was the total number of miles they drove in 2 days?

    _____

    d. It is 900.5 miles to the next place they plan to visit on their vacation. Benito's family wants to take two days to drive this distance. If they want to drive *more than* 345.8 miles on the third day, how far should they drive? How many miles will they need to drive on the fourth day?

    _____

1.04 Develop fluency with addition and subtraction of non-negative rational numbers with like denominators, including decimal fractions through hundredths. a) Develop and analyze strategies for adding and subtracting numbers.

Practice for the EOG Test — **Number & Operations Practice**

Name _____

NUMBER & OPERATIONS

1. Becky collected 53 eggs from the chicken coop on Monday. On Tuesday, she collected 95 eggs. For the next three days, she collected 47, 28, and 77 eggs. How many eggs did she collect in all?

   TIP What operation should you use to solve the problem?

2. Marco's Toys needs to mail 132 toys. The toys are put into 6 boxes, so that the same number of toys is in each box. How many toys are in 2 of the boxes?

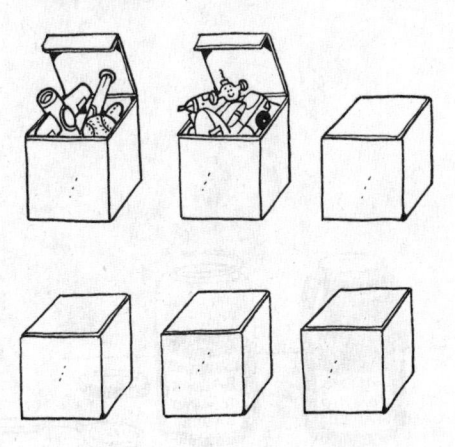

3. Tyesha is shopping. She wants to buy 4 pairs of pants in 4 different colors. The pants are $19.98 each with tax. She also wants to buy 4 matching shirts. They are on sale for 2 for $12.50 with tax. She does not want to spend more than $100.00. Estimate to find out if she can buy all eight items. Explain.

   Junior Shop
   PANTS 1 @ $19.98
   PANTS 1 @ $19.98
   PANTS 1 @ $19.98
   PANTS 1 @ $19.98
   SHIRT 2 @ $12.50
   SHIRT 2 @ $12.50
   Total   ?

4. **Explain It** The local cinema had unusually large crowds in the first two weeks of July. It sold 8,290 tickets in the first week and 2,010 tickets in the second week. How many tickets did it sell in all? How many *more* tickets did it sell in the first week than in the second week?

   Explain which method you used to compute: mental computation, estimation, calculator, or paper and pencil.

1.05 Develop flexibility in solving problems by selecting strategies and using mental computation, estimation, calculators or computers, and paper and pencil.

Number & Operations Practice   Practice for the EOG Test 37

Name _____

NUMBER & OPERATIONS

5. Sally sells homemade tomato sauce at the Farmers' Market. If she uses 15 tomatoes for $1\frac{1}{2}$ quarts of tomato sauce, how much tomato sauce can she make from 45 tomatoes?

6. Carly says 156 ÷ 12 is *more than* 144 ÷ 12. Is she correct? Without solving, how do you know?

7. Pay Mart received a large shipment of paper plates for their summer sale. They received 148 cases containing 30 packages in each case. How many packages of paper plates did they receive?

TIP How can you break apart 148 into numbers that are easier to compute and then solve a simpler problem?

8. **Explain It** A film processing plant developed 2,083 rolls of 24-exposure film over the weekend. How many pictures did it develop?

Explain which method you used to compute: mental computation, estimation, calculator, or paper and pencil.

1.05 Develop flexibility in solving problems by selecting strategies and using mental computation, estimation, calculators or computers, and paper and pencil.

38 Practice for the EOG Test

Number & Operations Practice

Name _____

NUMBER & OPERATIONS

9. Joey listed some of the heights of the tallest mountain peaks in the United States.
   - Sassafras Mountain, in South Carolina, is 3,560 feet tall.
   - Spruce Knob, in West Virginia, is 4,863 feet tall.
   - Mt. Mitchell, in North Carolina, is 6,684 feet tall.
   - Clingmans Dome, in Tennessee, is 6,643 feet tall.
   - Humphrey's Peak, in Arizona, is 12,633 feet tall.
   - Black Mountain, in Kentucky, is 4,145 feet tall.
   - Taum Sauk Mountain, in Missouri, is 1,772 feet tall.
   - Mt. Hood, in Oregon, is 11,239 feet tall.

Some of the Tallest Mountain Peaks in the United States

State	Peak	Elevation (in feet)

a. He wants to organize the peaks by height in a table. Complete Joey's table. Explain how you organized the heights.

_____

b. Which peak is the tallest? Which peak is the lowest? What is the range of the data?

_____

c. Which two peaks are *less than* 100 feet apart in height? How far apart are they?

_____

d. Brasstown Bald, in Georgia, is 4,786 feet tall. If Brasstown Bald were added to the table, how would the new table look? Describe the new table.

_____

1.05 Develop flexibility in solving problems by selecting strategies and using mental computation, estimation, calculators or computers, and paper and pencil.

Number & Operations Practice — Practice for the EOG Test

Name _____

NUMBER & OPERATIONS

10. Angie and Melissa took their puppies to the vet. Each puppy was weighed. Melissa's puppy weighed 2 pounds less than twice the weight of Angie's puppy. Together the 2 puppies weigh 43 pounds.

   2 pounds

   Melissa's puppy | ? pounds | ? pounds | |
   Angie's puppy | ? pounds |
   } 43 pounds

   a. Angie and Melissa drew this diagram to help them find how much each puppy weighs. How much does each puppy weigh?

   _____

   b. Write a problem that can be solved using the diagram below. Label the diagram to show the relationships in your problem. Then solve your problem. Show your work.

   TIP Look at the diagram. How do the parts shown in the diagram relate to each other?

   _____
   _____
   _____

1.05 Develop flexibility in solving problems by selecting strategies and using mental computation, estimation, calculators or computers, and paper and pencil.

40 Practice for the EOG Test   Number & Operations Practice

Name _____

NUMBER & OPERATIONS

1. There are sixty-three thousand, three hundred sixty inches in a mile. How else can this number be written?

   A 6,336
   B 63,306
   C 63,360
   D 63,366

2. Jake is buying 5 pounds of ground meat for $2.99 a pound and 5 packages of buns for $2.15 each. If he pays with a $50 bill, *about* how much change should he receive?

   A $15
   B $25
   C $35
   D $45

3. Look at the models.

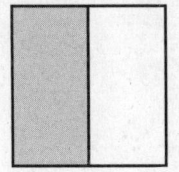

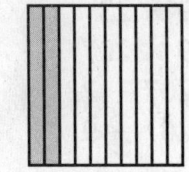

   Which of the following is a true statement shown by the models?

   A $\frac{1}{2} = 0.2$
   B $\frac{1}{2} < 0.2$
   C $\frac{1}{2} > 0.2$
   D $\frac{1}{2} = 0.02$

4. The fourth graders set up chairs for the school pep rally. There were 15 rows, with 30 chairs in each row. How many chairs were there in all?

   A 2
   B 45
   C 450
   D 4,500

5. Mark and Willa shared a pizza. Mark ate $\frac{1}{8}$ of the pizza, and Willa ate $\frac{5}{8}$ of the pizza. How much of the pizza did they eat in all?

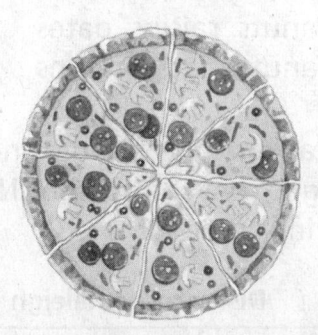

   A $\frac{1}{8}$
   B $\frac{1}{4}$
   C $\frac{1}{2}$
   D $\frac{3}{4}$

6. Robyn used three pieces of ribbon to make party decorations. The pieces were 2.8 meters, 3.1 meters, and 4.9 meters long. *About* how much ribbon did she use in all?

   A 6 meters
   B 9 meters
   C 11 meters
   D 13 meters

Go to next page ▶

Number & Operations Practice Test   **Practice for the EOG Test 41**

Name _____

NUMBER & OPERATIONS

7. Look at the recipe for Trail Mix.

Trail Mix
$\frac{1}{2}$ cup dates
$\frac{1}{3}$ cup peanuts
$\frac{3}{4}$ cup raisins

Which shows the ingredients in order from the *least* amount to the *greatest* amount?

A dates, peanuts, raisins
B dates, raisins, peanuts
C peanuts, raisins, dates
D peanuts, dates, raisins

8. The table shows the distances to Raleigh from some other North Carolina cities.

Distances to Raleigh

City	Distance (in miles)
Asheville	248
Charlotte	169
Nags Head	196

Maria drove from Asheville to Raleigh at an average speed of 62 miles per hour. How long was the trip?

A 2 hours
B 4 hours
C 6 hours
D 8 hours

9. Hal's video game score was 2,301. To find Anna's score, add twice the value of the hundreds digit to Hal's score. What was Anna's score?

A 2,307
B 2,361
C 2,901
D 6,301

10. Al solved this division problem.

```
     23
  9)207
    -18
     27
    -27
      0
```

Which computation could he use to check his answer?

A 9×23
B 9×207
C $207 + 23$
D $207 + 9$

11. The fourth graders are boxing groceries for the local food bank. They put 20 cans in each box. They have packed 60 boxes so far. How many cans have they packed?

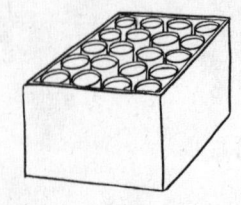

A 12
B 120
C 1,200
D 12,000

Go to next page ▶

42 Practice for the EOG Test   Number & Operations Practice Test

Name _____

NUMBER & OPERATIONS

12. Jamal is reading a 423-page book. If he reads 11 pages per day, *about* how many days will it take him to finish the book?

    A 4
    B 10
    C 42
    D 420

13. Each weekday morning, Bob drives to Chicago from Arlington Heights. On Monday evenings, he drives to Roselle for volunteer work. Then he drives 12.4 miles home. On Tuesday, he drives straight home. Which is a reasonable estimate of the number of miles Bob drives on Monday and Tuesday?

    Driving Distances

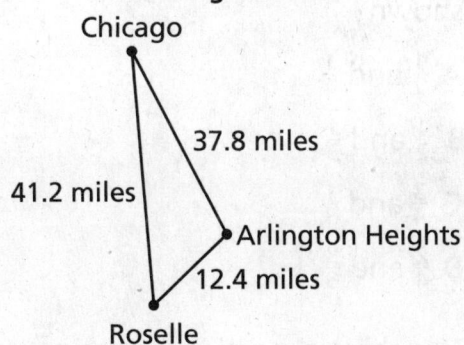

    Chicago
    37.8 miles
    41.2 miles
    Arlington Heights
    12.4 miles
    Roselle

    A 176 miles
    B 167 miles
    C 152 miles
    D 102 miles

14. The snack bar sells about 7 boxes of granola bars during each baseball game. There are 18 bars in a box. *About* how many bars will be sold during 2 games?

    A more than 300
    B between 200 and 300
    C between 100 and 200
    D fewer than 100

15. Sharon is making quilt blocks. Some triangles are white and some triangles are shaded. Of the blocks shaded, what mixed number is shown?

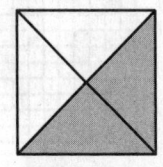

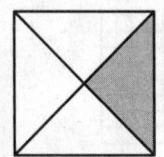

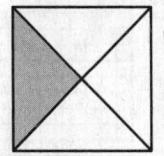

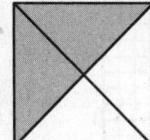

    A $1\frac{1}{2}$
    B $3\frac{1}{2}$
    C $3\frac{3}{4}$
    D $4\frac{1}{4}$

16. Abu ordered 20 packs of neon labels. Each pack contains 16 labels. Which operations can Abu use in the circles to find how many labels he ordered altogether?

    20 ◯ 16 = (20 × 10) ◯ (20 × 6)

    A × and +
    B × and ×
    C + and −
    D + and ÷

17. Salli has a board that is $\frac{11}{12}$ foot long. She cut a $\frac{5}{12}$-foot piece to make the floor of her birdhouse. How long is the piece of the board that is left?

    A $\frac{4}{3}$ feet
    B $\frac{7}{12}$ foot
    C $\frac{1}{2}$ foot
    D $\frac{5}{12}$ foot

    Go to next page ▶

Number & Operations Practice Test

18. Sharlene says, "I have twenty-one hundredths of a dollar. That's $0.21." Which model shows that amount?

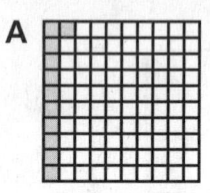

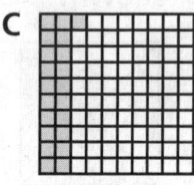

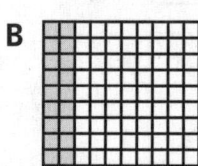

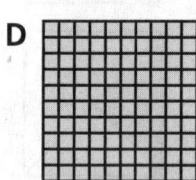

A
B
C
D

19. A bank has 5,000 pennies to be put into rolls of 50 pennies each. How many rolls will it have?

Number of pennies	50	500	5,000
Number of rolls	1	10	■

A 10
B 100
C 1,000
D 10,000

20. Harlan measured the amount of dried dog food he fed his dog, King. He put $2\frac{3}{8}$ cups of dried dog food in King's bowl. When King finished eating, there was $\frac{2}{8}$ cup left. How much dried dog food did King eat?

A $2\frac{5}{8}$ cups
B $2\frac{3}{8}$ cups
C $2\frac{1}{8}$ cups
D $\frac{2}{8}$ cup

21. Pocatello, Idaho, is 504 miles from Lewiston, Idaho. If the average speed of travel is 55 miles per hour, *about* how many total hours would it take to drive to Lewiston from Pocatello?

A 9 hours
B 12 hours
C 30 hours
D 60 hours

22. Look at the models.

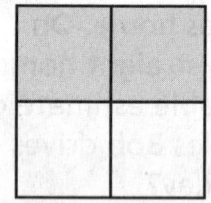

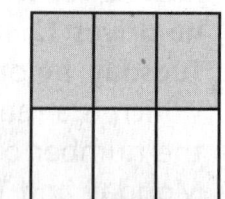

Which equivalent fractions are shown?

A $\frac{1}{2}$ and $\frac{1}{4}$
B $\frac{1}{2}$ and $\frac{1}{3}$
C $\frac{2}{4}$ and $\frac{1}{3}$
D $\frac{2}{4}$ and $\frac{3}{6}$

23. The shaded portion of the model shows the length in decimeters of an Acteon beetle, which is the bulkiest insect in the world. Which decimal shows the length of an Acteon beetle?

A 1.90 decimeters
B 0.90 decimeters
C 0.10 decimeters
D 0.09 decimeters

Go to next page ▶

Name _____

NUMBER & OPERATIONS

24. 12 × 11 is *more than* 11 × 11. How much more?

   A 11   C 121
   B 12   D 144

25. Mario had two paper chains. He taped the paper chains together. The first chain was $3\frac{2}{10}$ feet long. Now his paper chain is $7\frac{5}{10}$ feet long. How long was the second paper chain?

   A $10\frac{7}{10}$ feet

   B $10\frac{1}{2}$ feet

   C $4\frac{7}{10}$ feet

   D $4\frac{3}{10}$ feet

26. Bruce is visiting the Jungle Garden. He would like to see two shows. He will be at the Jungle Garden from 1:00 to 6:15, and he wants to see one of the new shows.

   Jungle Garden's Show Schedule

Show	Times
Parrot	1:20–3:30; 3:30–5:40
Sea Lion NEW	1:20–3:40; 3:50–6:10
Dolphin	3:30–5:35; 4:20–6:25
Safari Joe NEW	1:40–4:00; 4:00–6:20

   Which two shows can he see?

   A Sea Lion, then Dolphin
   B Sea Lion, then Parrot
   C Parrot, then Sea Lion
   D Safari Joe, then Dolphin

27. Rinne pool has an area of 22.32 square yards. The new Springs pool has an area of 35.41 square yards. How much larger is the new Springs pool?

   A 13.07 square yards
   B 13.09 square yards
   C 53.19 square yards
   D 57.73 square yards

28. Jeff wants to buy an adult sweatshirt and an adult baseball cap. He budgeted $60 for souvenirs. Does he have enough money? If not, how much more does he need? Decide whether you need an exact answer or an estimate.

   Souvenir Price List

Item	Price
Cap/adult	$24
Cap/child	$17
Sweatshirt/adult	$49
Sweatshirt/child	$32

   A yes; exact
   B no; $13 more; exact
   C yes; estimate
   D no; $20 more; estimate

29. Poundcakes Bakery slices each loaf of bread into 20 slices. For the morning rush, Sherry has sliced 80 loaves of bread. How many slices of bread is that?

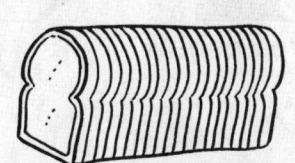

   A 16,000   C 160
   B 1,600   D 16

Go to next page ▶

Number & Operations Practice Test   Practice for the EOG Test 45

Name _____

NUMBER & OPERATIONS

30. The Kellys bought two sleeping bags. The Comfortplus cost $94.93 and the Overnighter cost $59.95. *About* how much more did they spend on the Comfortplus sleeping bag than on the Overnighter?

    A $35
    B $60
    C $100
    D $155

31. Carlos drew this picture to show 5 times 6. Which of the following shows the related division equation?

    A 6 × 5 = 30
    B 30 ÷ 5 = 6
    C 30 ÷ 15 = 2
    D 2 × 15 = 30

32. Which is a *more* reasonable number of colored pencils in the full box?

    20 colored pencils

    A 50
    B 75
    C 100
    D 120

33. Which national park is the third largest in this table?

    National Parks

Park	Size (in acres)
Acadia	47,633
Apostle Islands	69,372
Badlands	64,000
Golden Gate	75,398

    A Acadia
    B Apostle Islands
    C Badlands
    D Golden Gate

34. Miriam is weighing rocks for a science project. One rock weighs 17.29 pounds. Which is the amount rounded to the nearest tenth?

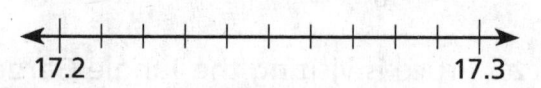

    17.2   17.3

    A 17 pounds
    B 17.2 pounds
    C 17.3 pounds
    D 20 pounds

35. A bead box has 30 beads in one compartment. *About* how many beads are in the entire bead box?

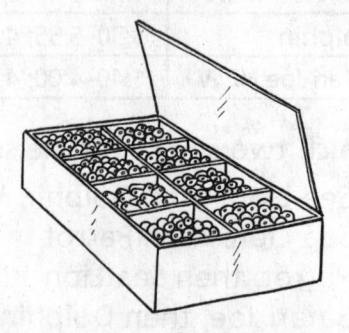

    A 60   C 120
    B 90   D 240

Go to next page ▶

46 Practice for the EOG Test   Number & Operations Practice Test

Name _____

NUMBER & OPERATIONS

36. In a relay race, Willis ran his leg of the race in 15.03 seconds. Thomas ran his leg in 12.89 seconds. Together, how many seconds did their portion of the race take?

   A 27.92 seconds
   B 27.82 seconds
   C 3.86 seconds
   D 2.14 seconds

37. Alvin parks his car in a parking garage each week. The table shows how much he paid each week in April to park.

   April Parking Fees

Week	Fee
1	$25.04
2	$31.93
3	$29.07
4	$24.72

   Which is the average amount he pays for 1 week?

   A $7.21
   B $15.83
   C $27.69
   D $110.76

38. The Garden Club is using a number pattern to determine the number of plants they can plant in a square shape. Which is the next number in the pattern?

   4   9   16   25

   A 36
   B 49
   C 64
   D 81

39. An amusement park has a theatre with 14 sections of seats. Each section has 5 rows of 19 seats. How many people can be seated for each show?

   A 70
   B 95
   C 266
   D 1,330

40. Alice ate $\frac{2}{3}$ cup of vanilla ice cream. Bill ate $\frac{1}{2}$ cup of chocolate ice cream. Rosa ate $\frac{3}{4}$ cup of butterscotch pudding. Order the fractions from *least* to *greatest*.

   A $\frac{3}{4}, \frac{2}{3}, \frac{1}{2}$ C $\frac{1}{2}, \frac{3}{4}, \frac{2}{3}$
   B $\frac{2}{3}, \frac{3}{4}, \frac{1}{2}$ D $\frac{1}{2}, \frac{2}{3}, \frac{3}{4}$

41. Mr. Garner is baking cookies. He put the first batch of 12 cookies onto paper to cool and went to get his mail. When he returned, there were only 9 cookies left. What fraction of the cookies were missing?

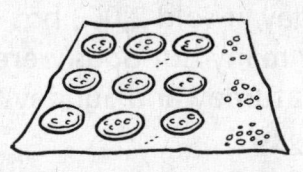

   A $\frac{3}{4}$ C $\frac{1}{4}$
   B $\frac{1}{2}$ D $\frac{1}{12}$

Go to next page ▶

Number & Operations Practice Test   Practice for the EOG Test

Name _____

NUMBER & OPERATIONS

42. Sam has 253 flowers to sell. He plans to put them in bunches of 3, each tied with a ribbon. *About* how many bunches will he have?

    A between 70 and 80
    B between 80 and 90
    C between 90 and 100
    D between 100 and 110

43. Janelle rides her bike 2.25 miles to her Aunt Betsy's house, then another 1.96 miles to her grandmother's house. After she visits with her grandmother, she rides 0.25 mile home. *About* how far does she ride her bike in all?

    A 4 miles
    B 3 miles
    C 2 miles
    D 1 mile

44. The Martin's Pet Store has 58 birds for sale. On Saturday, 7 customers bought 2 birds each. How many birds are left?

    A 44
    B 49
    C 102
    D 132

45. A minor league stadium sold 1,186 hot dogs at a Saturday game. On Sunday, it sold 2,063 hot dogs. How many hot dogs were sold on Saturday and Sunday?

    A 3,249
    B 3,219
    C 3,119
    D 877

46. A beluga whale in captivity eats about 511 pounds of fish per week. How many pounds per day is that?

    January

Sun	Mon	Tue	Wed	Thu	Fri	Sat
				1	2	3
4	5	6	7	8	9	10

    A 73 pounds
    B 504 pounds
    C 518 pounds
    D 703 pounds

47. Joey and Sam want to record their favorite band music on a CD. One song is 36 minutes and the other song is 19 minutes. They want to use mental computation to find the total number of minutes of the songs. Which number sentence could they use?

    A 40 + 20
    B 39 + 20
    C 38 + 20
    D 35 + 20

48. The table shows the number of color copies, *c*, a color printer prints per minute, *m*. How many copies can be printed in 9 minutes?

Input	m	4	5	6	7
Output	c	24	30	36	42

    A 42
    B 49
    C 54
    D 63

STOP

48 Practice for the EOG Test

Number & Operations Practice Test

Name _____

MEASUREMENT

Vocabulary

Circle the term that describes all of the other terms. Explain how the other terms in the list are related.

1. decade
   minute
   time
   second

   _____
   _____

2. ounce
   weight
   pound
   ton

   _____
   _____

3. inch
   mile
   length
   centimeter
   kilometer

   _____
   _____

4. pint
   capacity
   cup
   gallon
   liter

   _____
   _____

Answer each question for each unit of measure. Write *yes* or *no* in the table.

	Longer than your thumb?	Wider than your hand?	Longer than the distance from home to school?	Longer than your math book?
inch				
foot				
yard				
mile				
millimeter				
centimeter				
meter				
kilometer				

Measurement Vocabulary Practice   Practice for the EOG Test   49

Name _____

MEASUREMENT

Vocabulary

Tell whether each statement is *reasonable* or *not reasonable*. If the statement is not reasonable, change the underlined unit of measure so that the statement makes sense.

1. The temperature on a warm, summer day was 80 °C.

2. It took Jaime about 15 seconds to read a chapter in his book.

3. Helga's grandfather has been alive for over 7 decades.

4. Raul's dog is about 20 feet long.

5. The football field is 100 yards long.

6. Max drinks about 8 quarts of water each day.

7. Since Vinh drinks a lot of milk, his mother buys 2 pints of milk each week.

8. Deon's hamster weighs about 4 ounces.

9. An adult African elephant can weigh as much as 7 pounds.

10. Kayla's fingernail is about 8 centimeters across.

11. Alex brought 3 liters of juice to share with his classmates.

12. An apple has a mass of about 100 kilograms.

1. _____
2. _____
3. _____
4. _____
5. _____
6. _____
7. _____
8. _____
9. _____
10. _____
11. _____
12. _____

For 13–15, use the figure at the right. Complete each sentence with a term from the box.

area
feet
lengths
perimeter

13. The _____ of the rectangle is 32 feet.

14. The _____ of the rectangle is 60 square feet.

15. To find the perimeter of a polygon, you find the sum of the _____ of the sides.

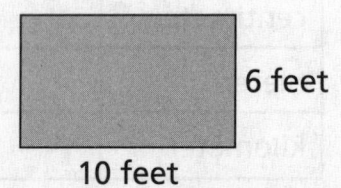

6 feet
10 feet

50 Practice for the EOG Test

Measurement Vocabulary Practice

Name _____

MEASUREMENT

1. Manchu leaves for school at 7:47 A.M. What is another way to read or write that time?

TIP When you speak casually, how do you say the time?

_____
_____
_____

2. The first dolphin show at the Brookfield Zoo begins at 11:30 A.M. The last show begins at 4:00 P.M. and ends at 4:30 P.M. How much time elapses from the beginning of the first show to the end of the last show of the day?

Next Dolphin Show Starts at

_____
_____
_____

3. Melissa needs 2 yards of fabric to make a costume. The fabric bolt has 52 inches left on it. Is there enough fabric on the bolt for Melissa to make the costume? Explain.

_____
_____
_____

4. **Explain It** Alonzo is making punch for the class party. The recipe calls for 4 pints of orange juice. If he doubles the recipe, how many cups of orange juice will he need to measure? Explain how you found your answer.

Pint	1	2	3
Cup	2	4	6

_____
_____
_____

Maintains (3) 2.01 Solve problems using measurement concepts and procedures involving: a) Elapsed time. b) Equivalent measures within the same measurement system.

Measurement Practice   Practice for the EOG Test **51**

Name _____

MEASUREMENT

1. Alison is weighing a ball of yarn. She is deciding which scale to use. Should she use the bathroom scale or the food scale? Explain.

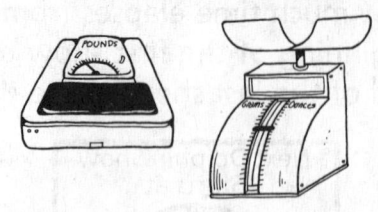

   _____
   _____
   _____
   _____

2. The maximum gross weight of tractor-semitrailer vehicles allowed on Chicago expressways is 88,000 pounds. How many tons is that?

   TIP How can you use multiples of 10 to change pounds to tons?

   _____
   _____
   _____
   _____

3. Esther is at the store buying juice for her party. Her punch recipe calls for 1 gallon of apple juice. The Pic-n-Go only sells liter containers of apple juice. Esther knows that 1 liter is a little more than 1 quart, so she estimates that she will need about 8 liters of apple juice. Is this a reasonable estimate? Will she have enough apple juice for her recipe? Explain.

   _____
   _____
   _____
   _____

4. **Explain It** Ralph and his brother measured their backyard with a meterstick. One side measured 20 meters. They wanted to know how long the side is in centimeters, so they divided by 10. They said the one side is 2 centimeters long. Are they correct? If not, describe their error and write the correct length.

   _____
   _____
   _____
   _____
   _____
   _____

Maintains (3) 2.01 Solve problems using measurement concepts and procedures involving: b) Equivalent measures within the same measurement system. **Maintains (3) 2.02** Estimate and measure using appropriate units. a) Capacity (cups, pints, quarts, gallons, liters). b) Length (miles, kilometers). c) Mass (ounces, pounds, grams, kilograms).

52 Practice for the EOG Test

Measurement Practice

Name _____

MEASUREMENT

1. Renee has a basket attached to the front of her bike. She wants to attach a ribbon around the perimeter of her basket. She drew a diagram to help her find the perimeter. What is the perimeter of the basket? *About* how long should her ribbon be?

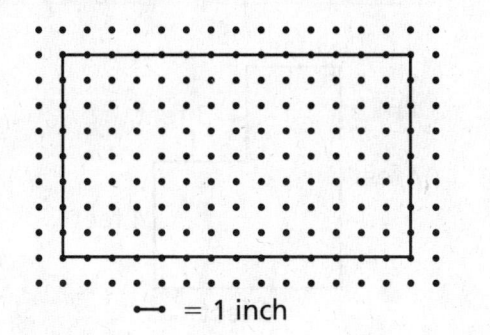

   ⟵⟶ = 1 inch

   _____
   _____

2. Mr. Miller drew this equilateral triangle on the board. It has a perimeter of 27 inches. What formula can you use to find the length of the unknown side? What is the unknown length?

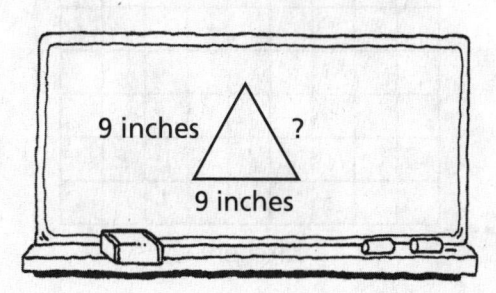

   _____
   _____

3. Hector drew this square to use as a pattern to make a quilt block. Use a metric ruler to find the perimeter of the square.

   _____

   If Hector makes a quilt block 5 squares by 5 squares, what is the perimeter of his quilt block?

   _____

4. **Explain It** Draw a rectangle with a perimeter of 12 units. Then record the lengths of the sides. How can you check your drawing?

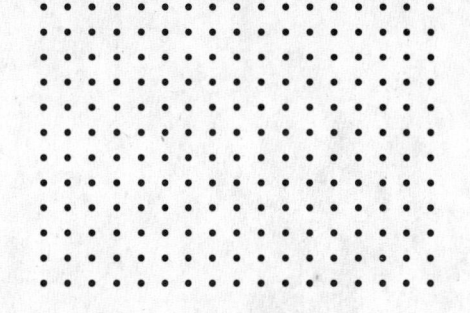

   _____
   _____
   _____

2.01 Develop strategies to determine the area of rectangles and the perimeter of plane figures.

Measurement Practice   Practice for the EOG Test 53

Name _____

MEASUREMENT

5. Kelly is helping his dad lay 1-square-foot tiles on the kitchen floor. What is the area of the kitchen?

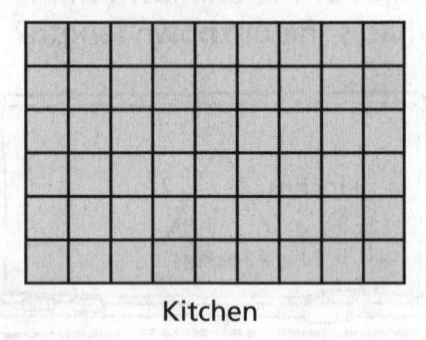

Kitchen

_____

6. Carrie's sandbox is shaped like an L. What is the area of the sandbox? Show one way you can divide it into rectangles so that you can find the area.

TIP How many rectangles make up the sandbox?

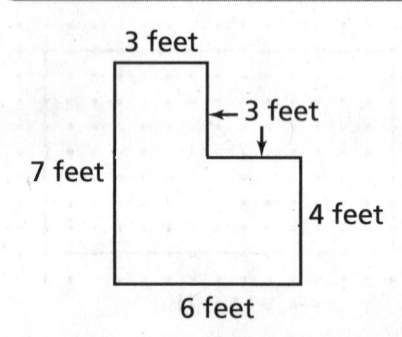

3 feet
← 3 feet
7 feet
4 feet
6 feet

_____

7. Sandy drew two figures that have different perimeters but the same area. Both of her figures have an area of 16 square units. Draw two figures that Sandy could have drawn. Find the perimeter of each figure.

8. **Explain It** The Pines Park kiddie pool is shown on the grid below. Each square equals 1 square foot. What is the *best* estimate of the pool's area? Explain your answer.

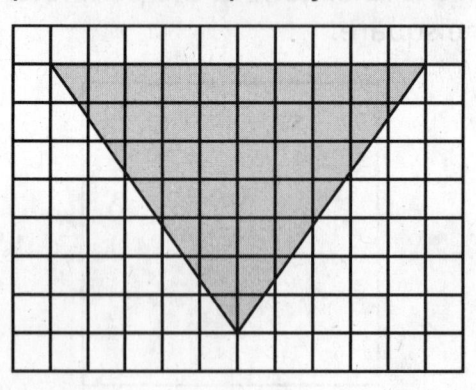

_____
_____
_____
_____

2.01 Develop strategies to determine the area of rectangles and the perimeter of plane figures.

54 Practice for the EOG Test

Measurement Practice

Name _____

MEASUREMENT

9. Michelle wants to build a patio. She drew these diagrams of possible shapes for her patio.

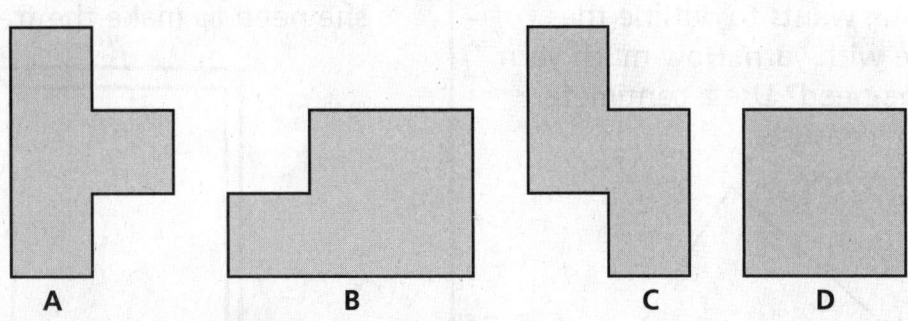

A B C D

a. Which figure has the *greatest* area? Explain how you know.

_____
_____
_____
_____

b. Which figure has the *least* perimeter? Explain how you know.

_____
_____
_____
_____

c. Draw a different figure that has the same area as Figure B.

	PS	Comp	Comm
IV			
III			
II			
I			

2.01 Develop strategies to determine the area of rectangles and the perimeter of plane figures.

Measurement Practice Practice for the EOG Test 55

Name _____

MEASUREMENT

1. Kim drew the triangle below on a greeting card she made for her sister. She wants to outline the triangle with yarn. How much yarn does she need? Use a centimeter ruler to measure.

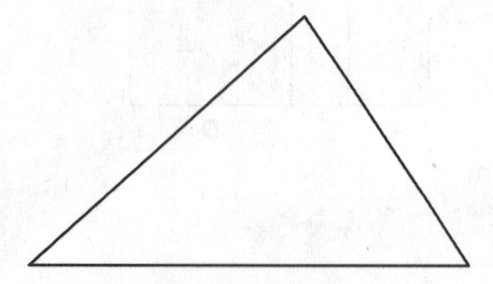

_____

2. Nancy drew a diagram of a wooden picture frame. How much wood will she need to make the frame?

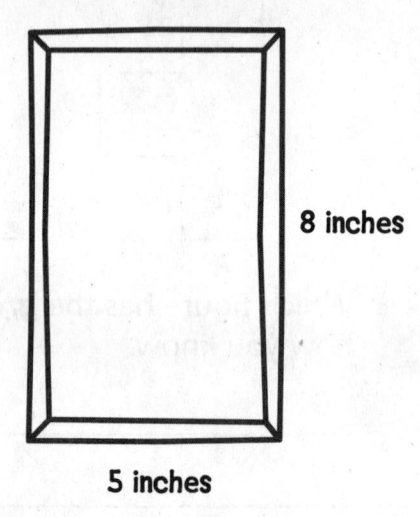

8 inches

5 inches

_____

3. Clay drew this diagram of the pen he built for his dog.

8 feet

Clay's dog pen has a perimeter of 40 feet. What is the length of the dog pen?

_____

4. **Explain It** The perimeter of the square equals the perimeter of the triangle. How long is each side of the square? Explain how you know.

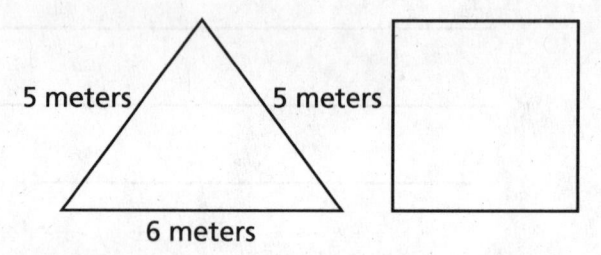

5 meters   5 meters

6 meters

_____
_____
_____
_____
_____
_____

2.02 Solve problems involving perimeter of plane figures and areas of rectangles.

56 Practice for the EOG Test

Measurement Practice

Name _____

MEASUREMENT

5. Jessie drew a diagram of the closet floor he will tile. Each square is 1 square foot. What is the area of the floor?

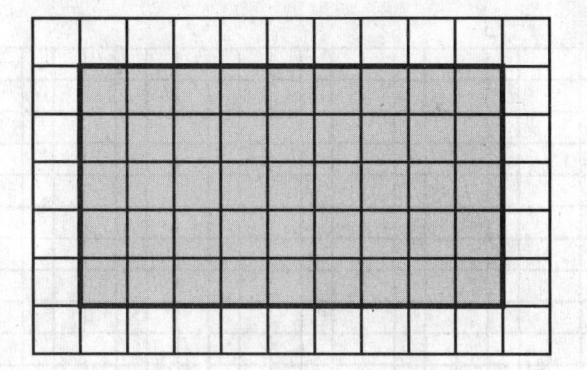

_____

6. Valerie wants to plant 12 daisy plants in a rectangular garden. She will allow 1 square foot of space for each plant. Draw a diagram of a rectangle that could be the shape of her garden. Label the length and width of your rectangle.

7. The band director had a special stage built for school band performances. What is the area of this stage? Show your work using pictures, words, or numbers.

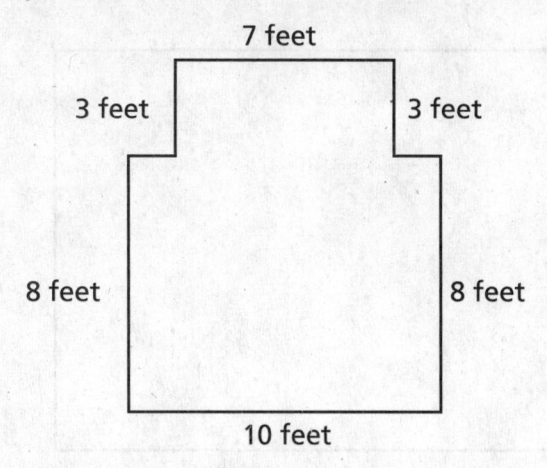

7 feet
3 feet   3 feet
8 feet   8 feet
10 feet

8. **Explain It** Tara made a quilt of small fabric squares. Each square is 4 inches by 4 inches. The quilt is 9 squares wide by 12 squares long. What is the area of the quilt? Explain how you found your answer.

TIP How can you determine the width and the length of the quilt?

_____
_____
_____
_____

2.02 Solve problems involving perimeter of plane figures and areas of rectangles.

Measurement Practice   Practice for the EOG Test 57

Name _____

MEASUREMENT

9. Matt wants to plant a rectangular garden with an area of 30 square feet.

   a. Draw three different rectangles that have an area of 30 square feet.

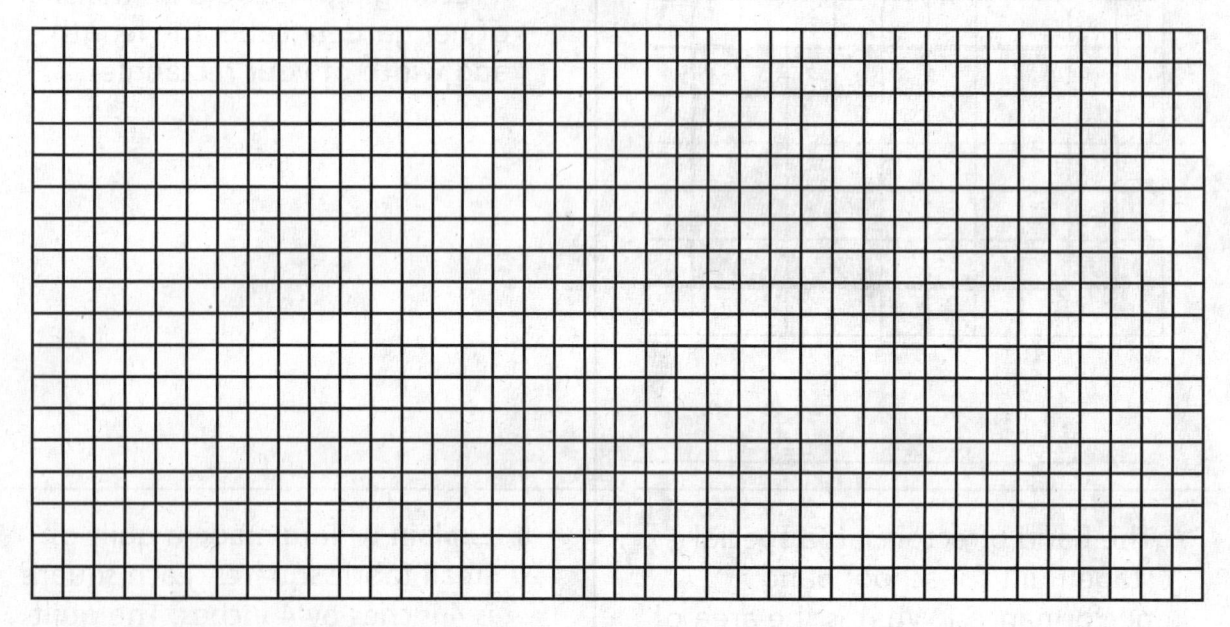

   b. Make a table to show the width, length, perimeter, and area of each rectangle you drew.

   c. For which rectangle would Matt need to buy the *least* amount of fencing to enclose the garden? Explain how you know.

   _____

   _____

   _____

2.02 Solve problems involving perimeter of plane figures and areas of rectangles.

58 Practice for the EOG Test   Measurement Practice

Name _____

MEASUREMENT

1. The Murphys have a pond in their garden. Each square measures 1 square meter. What is the **best** estimate of the area of the pond?

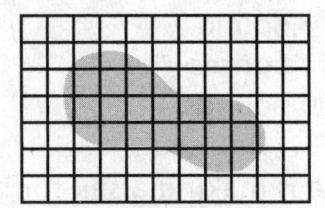

   A $18\frac{1}{2}$ square meters
   B $27\frac{1}{2}$ square meters
   C 45 square meters
   D 77 square meters

2. Which is the **best** estimate for the length of a new pencil?

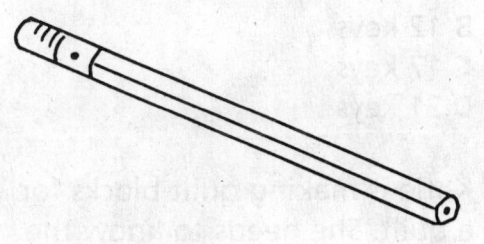

   A 8 centimeters
   B 18 centimeters
   C 8 meters
   D 18 meters

3. Maddie made a rectangle on a geoboard. What is its perimeter?

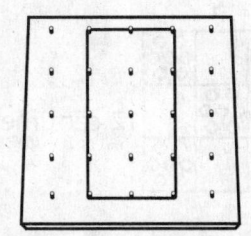

   A 8 units
   B 12 units
   C 15 units
   D 16 units

4. To measure enough water to make 2 gallons of lemonade, Maria is using a 2-cup measuring cup. How many times will she have to fill the measuring cup?

   A 32
   B 24
   C 16
   D 12

5. In preparation for buying new carpet for his bedroom, Jake is helping his dad find the area. What is the area of Jake's bedroom?

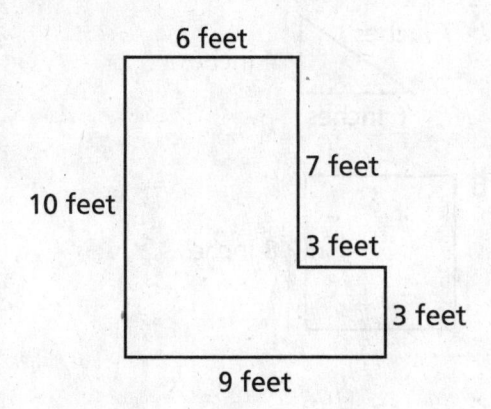

   A 38 square feet
   B 42 square feet
   C 60 square feet
   D 69 square feet

Go to next page ▶

Measurement Practice Test Practice for the EOG Test **59**

MEASUREMENT

6. Derek rode the train from Burlington to Raleigh. How long was his train ride?

NORTHBOUND TRAIN 74 SCHEDULE

City	Time
Greensboro	departs 7:10 P.M.
Burlington	departs 7:41 P.M.
Durham	departs 8:22 P.M.
Cary	departs 8:45 P.M.
Raleigh	arrives 9:10 P.M.

A 29 minutes
B 51 minutes
C 1 hour 29 minutes
D 1 hour 51 minutes

7. Which figure has the same perimeter as this regular hexagon?

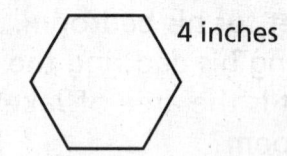

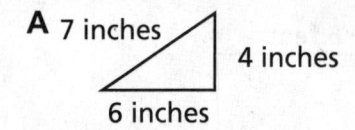

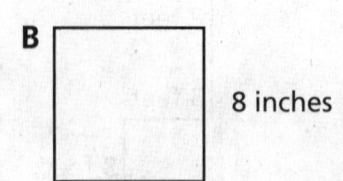

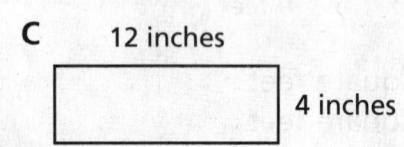

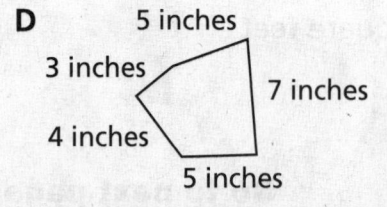

4 inches

A 7 inches, 4 inches, 6 inches
B 8 inches (square)
C 12 inches, 4 inches
D 5 inches, 3 inches, 7 inches, 4 inches, 5 inches

8. Which is the *most* reasonable measurement for the weight of a soccer ball?

A 1 ounce
B 1 pound
C 100 ounces
D 100 pounds

9. Tyree is making a sail for his model boat. He only has a key to measure his triangle. *About* how many keys around is the triangle?

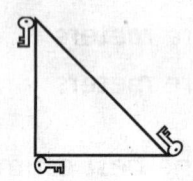

A 8 keys
B 12 keys
C 17 keys
D 21 keys

10. Katie is making quilt blocks for a quilt. She needs to know the length and width of each block to make sure that when all the quilt blocks are sewn together, they will cover her bed. The area of one quilt block is 144 square centimeters. What is the length of one quilt block?

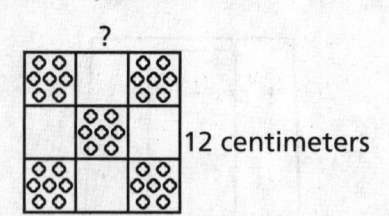

12 centimeters

A 48 centimeters
B 36 centimeters
C 24 centimeters
D 12 centimeters

Go to next page ▶

60 Practice for the EOG Test   Measurement Practice Test

Name _____

MEASUREMENT

11. Which of the following is the *shortest* length?

  A 132 inches
  B 10 feet 8 inches
  C 11 feet 2 inches
  D 1 yard 2 feet

12. The Johnsons have a new triangular patio. Its perimeter is 45 feet. One of its sides is 20 feet and another side is 15 feet. What is the length of the third side?

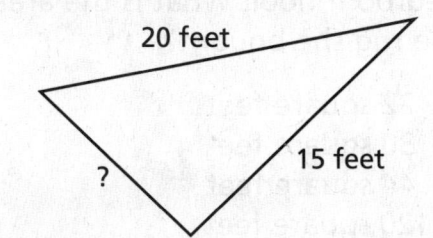

  A 35 feet
  B 25 feet
  C 20 feet
  D 10 feet

13. Walter grew a giant pumpkin that won a blue ribbon at the county fair. If the mass of the small pumpkin is 12 kilograms, *about* how many kilograms is the mass of Walter's pumpkin?

  A 60 grams
  B 600 grams
  C 60 kilograms
  D 600 kilograms

14. Hugh and his dad are making a backyard kennel for their dog. They drew four different designs for the kennel. Each design has the same perimeter. Which design has the *greatest* area?

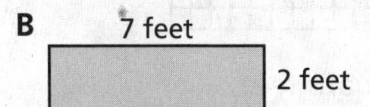

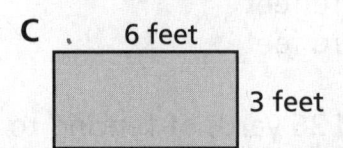

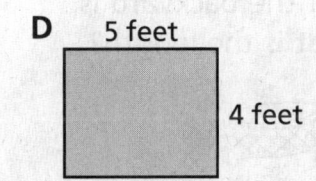

  A 8 feet × 1 foot
  B 7 feet × 2 feet
  C 6 feet × 3 feet
  D 5 feet × 4 feet

15. Alfredo has a soccer game at 3:05 P.M. If the game lasts 50 minutes, when will it be over?

  A 2:10 P.M.
  B 2:15 P.M.
  C 3:55 P.M.
  D 4:00 P.M.

Go to next page ▶

Measurement Practice Test   Practice for the EOG Test

Name _____

MEASUREMENT

16. Kaila drew a diagram of a pen she wants to build for her rabbit. Each square on the grid is 1 square foot. Which is the *best* estimate of the area of the pen?

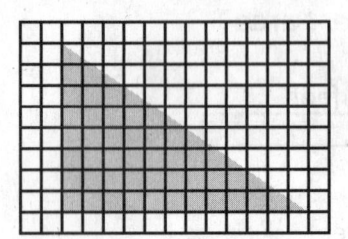

   A 30 square feet
   B 46 square feet
   C 51 square feet
   D 80 square feet

17. Scott used 36 yards of fencing to enclose his rectangular backyard. The width of the backyard is 6 yards. What is the length?

   A 9 yards   **C** 18 yards
   B 12 yards   **D** 24 yards

18. It is Sandy's job to fill and carry the water bucket to the campsite. *About* how much water does the bucket hold?

   A 8 milliliters   **C** 8 liters
   B 80 milliliters   **D** 80 liters

19. Marisa's bedroom floor is a rectangle that is 10 feet wide and 12 feet long.

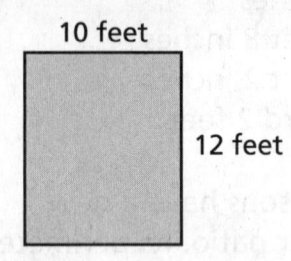

   10 feet
   12 feet

   She bought a rug that is half the length and half the width of the bedroom floor. What is the area of the rug she bought?

   A 22 square feet
   B 30 square feet
   C 44 square feet
   D 120 square feet

20. Mrs. Bryant is on a tight budget. She needs to buy the least amount of fencing for her garden. All four of the garden designs below have the same area. Which design has the *least* perimeter?

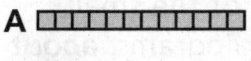

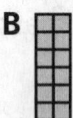

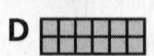

   A
   B
   C
   D

62 Practice for the EOG Test

Measurement Practice Test

Name _____

GEOMETRY

Vocabulary

Draw a polygon by connecting the lettered points on each coordinate grid. Then write the ordered pair for each point and name the plane figure.

1.

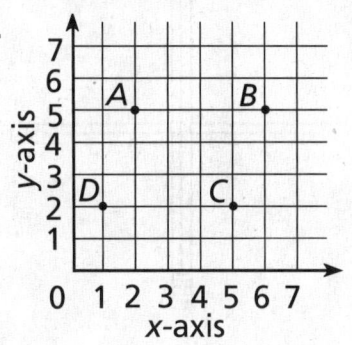

2.

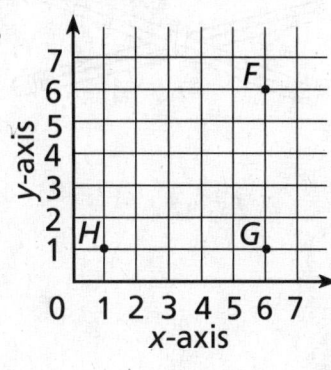

3.

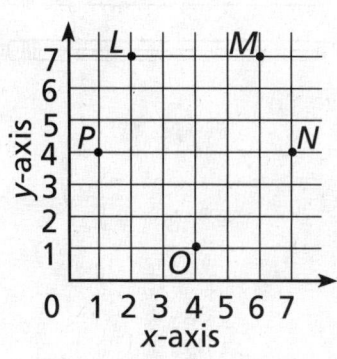

_____

_____

_____

_____

Use a coordinate grid to draw each polygon. Write the ordered pair for each point you used.

4. square

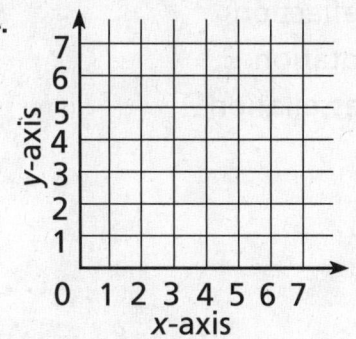

5. trapezoid

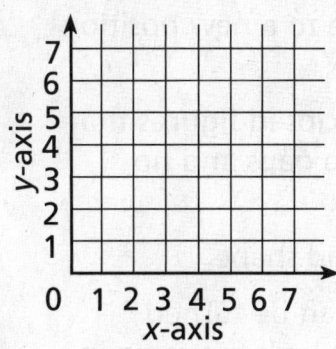

6. obtuse triangle

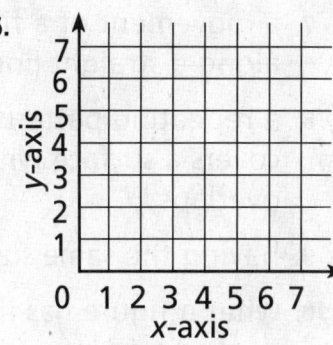

_____

_____

_____

_____

Geometry Vocabulary Practice   Practice for the EOG Test   63

Name _____

Vocabulary

Name the line relationships you see in each figure.
Write *intersecting*, *parallel*, or *perpendicular*.

1.

2.

3.

_____   _____   _____

_____   _____   _____

Write the letter of the term that matches each definition.

___ 4. a movement of a figure to a new position by turning the figure around a point

___ 5. having the same shape but possibly different in size

___ 6. what a figure has if it can be folded along a line so that its two parts match exactly

___ 7. a movement of a figure to a new position along a straight line

___ 8. a repeating pattern of closed figures that covers a surface with no gaps and no overlaps

___ 9. having the same size and shape

___ 10. what a figure has if it can be turned around a point and still look the same in at least two positions

___ 11. a movement of a figure to a new position by flipping it over a line

___ 12. the movement of a figure by a translation, reflection, or rotation

A. congruent
B. similar
C. line symmetry
D. rotational symmetry
E. transformation
F. translation
G. reflection
H. rotation
I. tessellation

Name _____

GEOMETRY

1. In a game, Curt has his ships hidden from view in these positions. What ordered pair names the position of the smallest ship?

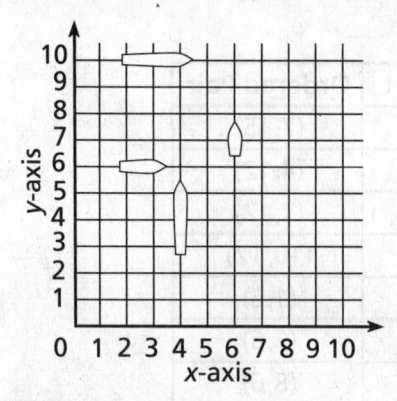

TIP What do the numbers in an ordered pair represent?

_____

2. Anna graphed the locations of her house, Lucy's house, and her aunt's house. Name the ordered pair that locates her aunt's house. Describe how Anna could go to her aunt's house from her house.

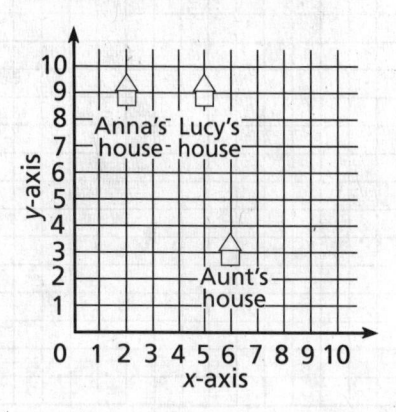

_____

3. On the coordinate grid, graph a polygon with vertices (2,1), (2,7), (8,7), and (8,1). Name the polygon. Then find the perimeter and the area of the plane figure.

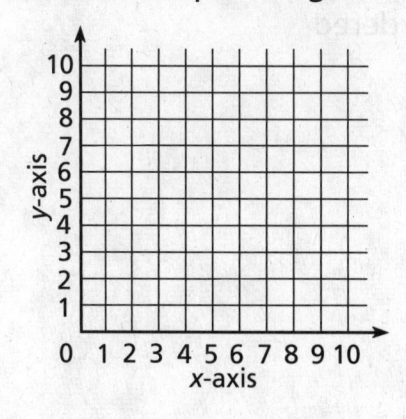

_____

4. **Explain It** Sonya said that her roses are planted at point (4,5). Is she correct? Explain why or why not.

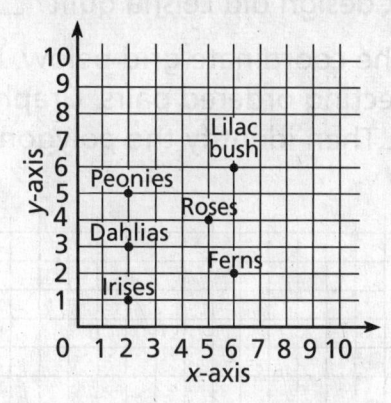

_____

3.01 Use the coordinate system to describe the location and relative position of points and draw figures in the first quadrant.

Geometry Practice Practice for the EOG Test **65**

Name _____

GEOMETRY

5. Leisha is designing a quilted seat pad for her kitchen chairs. After she sews the pieces together, she will quilt yarn knots along the outline of her design. Graph and label each ordered pair below to see where she plans to quilt the yarn knots.

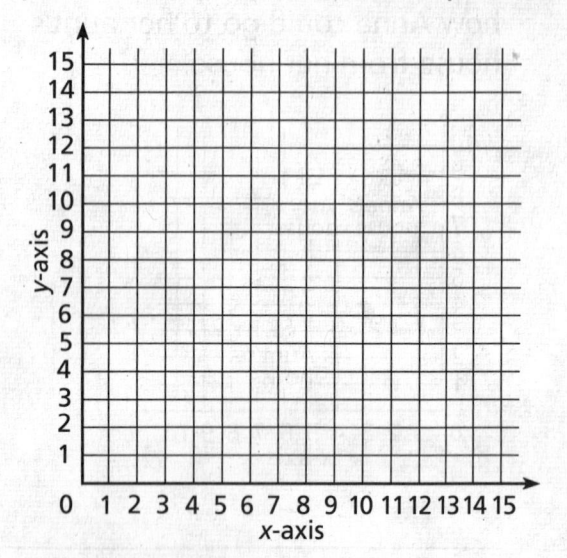

Point	Ordered Pair
A	(7,15)
B	(4,12)
C	(7,12)
D	(10,12)
E	(1,6)
F	(6,6)
G	(8,6)
H	(13,6)
I	(6,2)
J	(8,2)

a. Connect points A, B, and D. Name the polygon. _____

b. Connect points C, E, and H. Name the polygon. _____

c. Connect points F, G, J, and I. Name the polygon. _____

d. What design did Leisha quilt? _____

e. Use the coordinate grid below. Make a quilt design by connecting ordered pairs. Graph and label the ordered pairs. Then identify the polygons in your design.

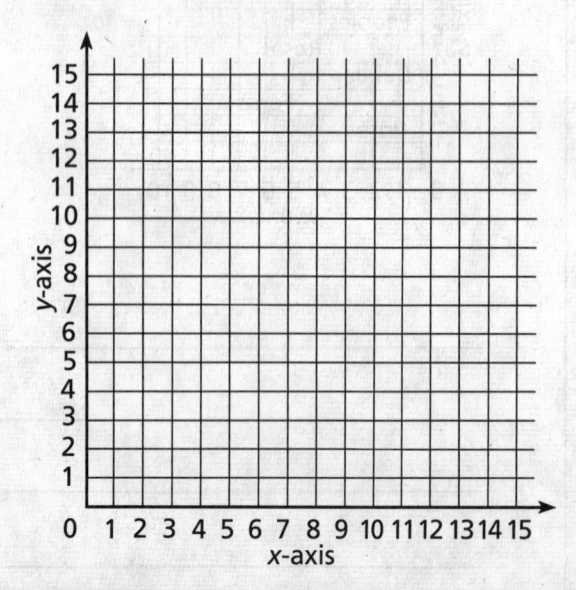

3.01 Use the coordinate system to describe the location and relative position of points and draw figures in the first quadrant.

66 Practice for the EOG Test   Geometry Practice

Name _____

GEOMETRY

1. Name any line relationship you see in the ropes of the swing. Write *intersecting*, *parallel*, or *perpendicular*.

   _____

2. Stephanie decorated the school bulletin board with letters. She said that all the letters she used appear to have perpendicular line segments. Explain her error.

   L T X
   E F H

   _____
   _____
   _____

3. Angela says her house is on the street that is perpendicular to Oak Street and the railroad tracks cross it. What street is her house on?

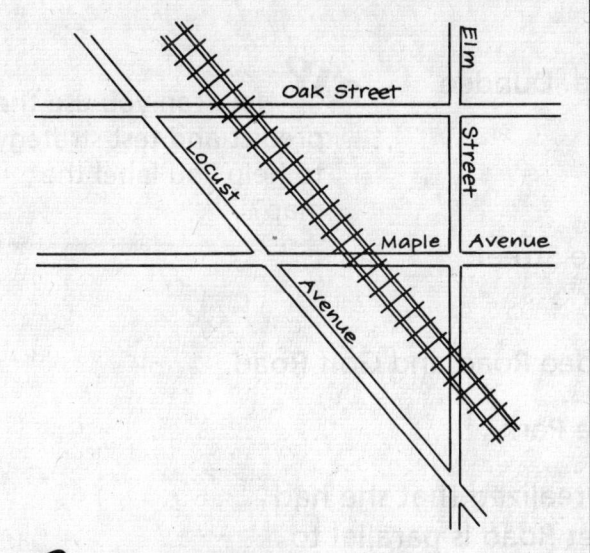

   TIP How can you use your understanding of line relationships to find your way around a map?

4. **Explain It** Aaron says that he sees three types of line relationships on a sheet of notebook paper. Describe the line relationships he sees.

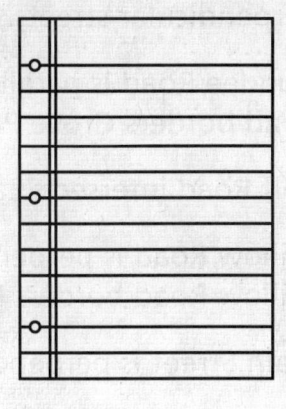

   _____
   _____
   _____
   _____
   _____
   _____

3.02 Describe the relative position of lines using concepts of parallelism and perpendicularity.

Geometry Practice   Practice for the EOG Test   67

Name _____

GEOMETRY

5. Caroline drew a map for a social studies assignment. Help her use the following descriptions to label the streets and parks on her map.

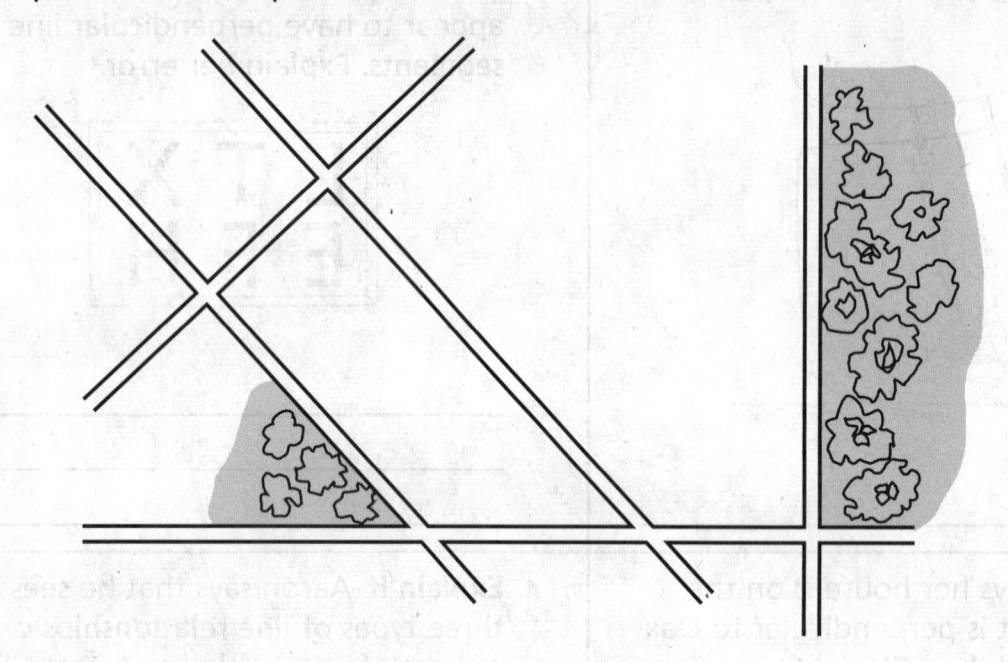

a. Klein Creek Park is located at a corner of two perpendicular streets.

b. Dundee Road is parallel to Golf Road. Dundee Road borders Grove Park.

c. Golf Road intersects Lake Street.

d. Willow Road is perpendicular to Lake Street. Willow Road borders Klein Creek Park.

e. Main Street is perpendicular to Dundee Road and Golf Road.

f. Two intersecting streets border Grove Park.

g. After Caroline labeled the map, she realized that she had forgotten to draw Picket Road. Picket Road is parallel to Dundee Road. Where might Picket Road be located on her map? Draw Picket Road on the map.

TIP How can you use the predict and test strategy to help you label the map?

_____
_____
_____

3.02 Describe the relative position of lines using concepts of parallelism and perpendicularity.

68 Practice for the EOG Test

Geometry Practice

Name _____

GEOMETRY

1. Yolanda wants to bake cookies. She has several cookie cutters to choose from. Which two cookie cutters have line symmetry?

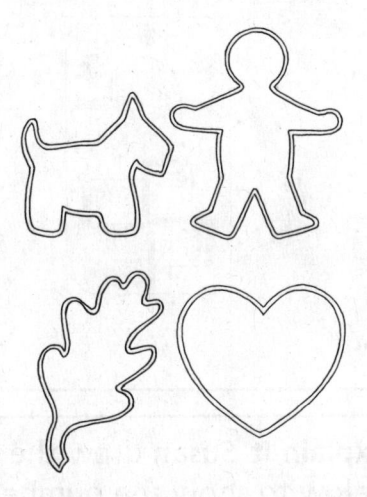

2. The state flower of North Carolina is the dogwood. Draw the lines of symmetry. Tell whether the dogwood flower has *line symmetry*, *rotational symmetry*, or *both*.

3. Vikki drew a square on 0.5-centimeter grid paper and another square on 1-centimeter grid paper. Both squares are 3 units on each side. Tell whether the two squares are *congruent*, *similar*, *both*, or *neither*. How do you know?

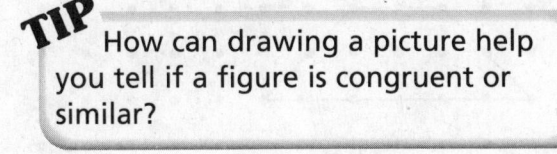

TIP How can drawing a picture help you tell if a figure is congruent or similar?

4. **Explain It** Dave is trying to find a missing puzzle piece. Identify the missing piece. Explain how you know.

A   B   C

Maintains (2) 3.03 Identify and make: a) Symmetric figures. b) Congruent figures.

Geometry Practice Practice for the EOG Test **69**

Name _____

GEOMETRY

1. The instruction sheet for Evelyn's bike shows a diagram of how to tighten the bolts on her bike. How many degrees should she turn the bolt? In what direction should she turn it?

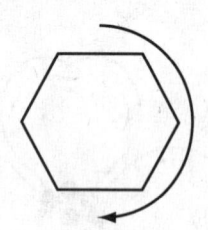

_____

2. Mark is making a poster for the school's Library Day. What transformation did he use to make his design?

_____

3. Jon is solving a puzzle. Predict how the piece needs to be moved to fit into the puzzle. Then test your prediction and describe the transformations Jon will need to use.

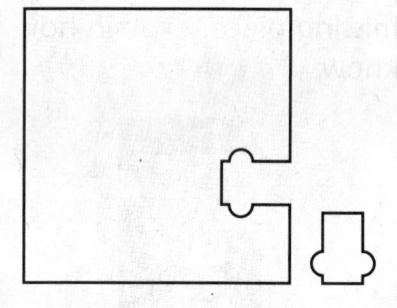

_____

4. **Explain It** Susan drew the diagram below to show the number of chairs that can be placed in the gym. She wants to know if there is room in the gym for 2 sections of chairs to be placed in the shape of a trapezoid next to each other. If she translates her pattern right, will there be room for 2 sections of chairs? Explain how you know.

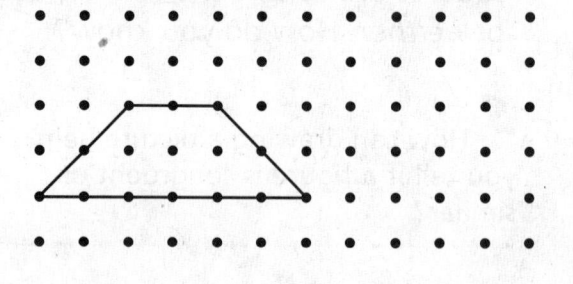

_____
_____
_____
_____

3.03 Identify, predict, and describe the results of transformations of plane figures. a) Reflections. b) Translations. c) Rotations.

70 Practice for the EOG Test   Geometry Practice

Name _____

GEOMETRY

5. A poster in Andy's room shows this design.

   What transformation was used to make the vase of flowers?
   _____

6. This piece of Japanese kimono fabric shows a tessellation. Describe the transformation used to move figure A to figure B.

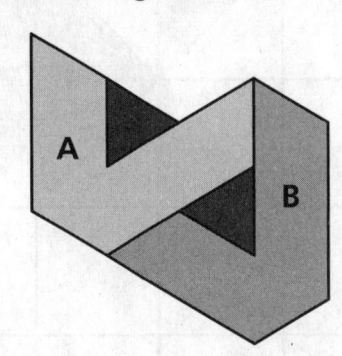

   _____

7. A town called Mitla in Mexico is famous for its stone mosaics. What transformation was used in this pattern on the wall? Write a rule for the pattern. Then draw the next two figures.

   TIP Was the pattern unit flipped, turned, or slid?

   _____

8. **Explain It** Sharon is designing stationary. Describe a rule that she could use to make her pattern. Describe how you can identify a rule for a geometric pattern.

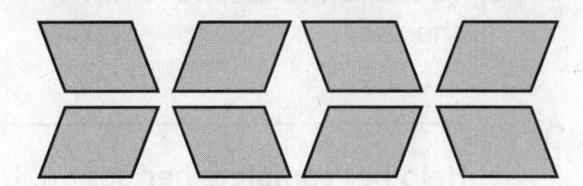

   _____

3.03 Identify, predict, and describe the results of transformations of plane figures. a) Reflections. b) Translations. c) Rotations.

Geometry Practice   Practice for the EOG Test   **71**

Name _____

GEOMETRY

9. Mrs. Cooley is making a wall hanging. She only completed half of the design. Help Mrs. Cooley finish her design.

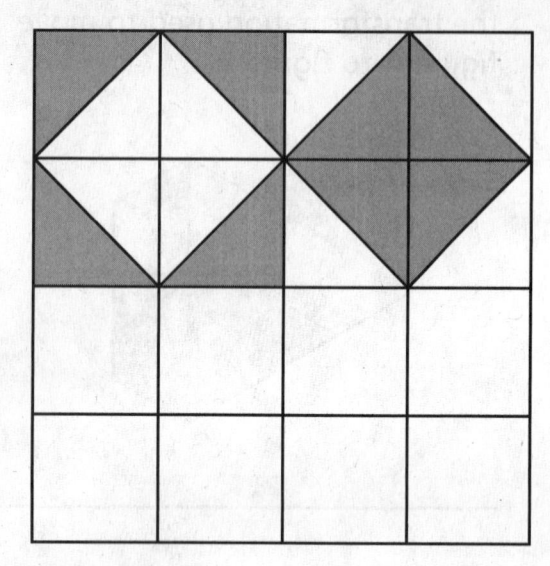

a. Mrs. Cooley used this figure to make her design. Describe the transformation she used to make the first row of her design.

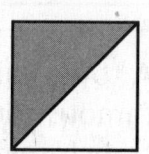

_____

b. To make the second row, what transformation did she use?

_____

c. Help her complete her design. Use this figure to complete the design. Describe the transformations you used.

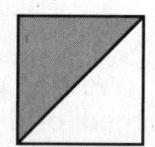

_____
_____
_____
_____

3.03 Identify, predict, and describe the results of transformations of plane figures. a) Reflections. b) Translations. c) Rotations.

72 Practice for the EOG Test

Geometry Practice

Name _____

GEOMETRY

1. Brian installed a picket fence around his yard. Which terms **best** identify the line relationships the fence models?

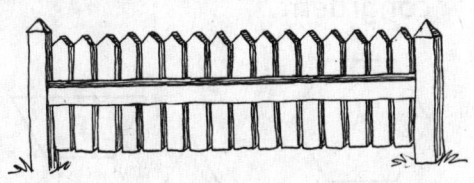

   A perpendicular and intersecting
   B parallel and intersecting
   C parallel, perpendicular, and intersecting
   D parallel

2. Amber's grandmother made a quilt with this pattern. Which describes how the triangle was transformed?

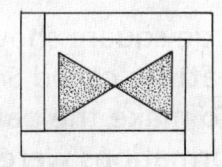

   A translation
   B reflection
   C 90° rotation
   D 90° counterclockwise rotation

3. Many shells have symmetry. Below is a chiton. Which of the following **best** describes the chiton?

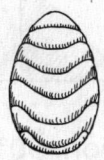

   A It has line and rotational symmetry.
   B It has rotational symmetry.
   C It has line symmetry.
   D It does not have line symmetry.

4. Allen moved the airplane game piece from position A to position B on the board. Name the transformation he used to move the airplane game piece.

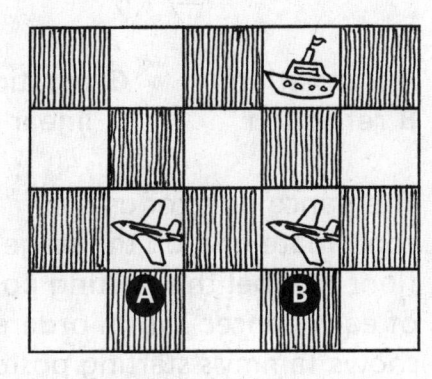

   A translation
   B reflection
   C 90° rotation
   D 180° rotation

5. Mauricio keeps his tools on a pegboard. The pegboard is labeled so that each tool is always put back in the same place. Which ordered pair shows where his wrench goes?

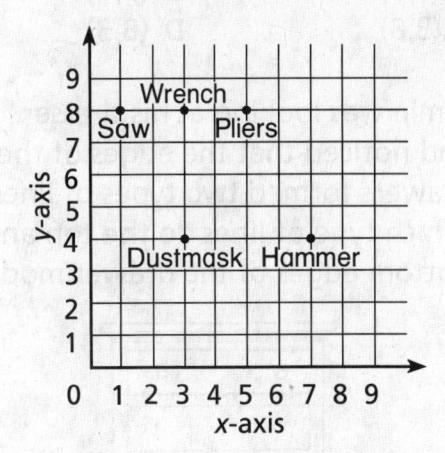

   A (7,4)
   B (3,8)
   C (8,3)
   D (1,8)

Go to next page ▶

Geometry Practice Test — Practice for the EOG Test 73

Name _____

GEOMETRY

6. This quilt pattern is named Wyoming Whirligigs. Which transformation does this pattern model?

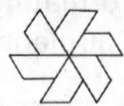

   A translation
   B reflection
   C rotation
   D linear

7. The dance director uses a coordinate grid on the stage dance floor to label the starting position of each dancer. Which ordered pair shows Tammy's starting position?

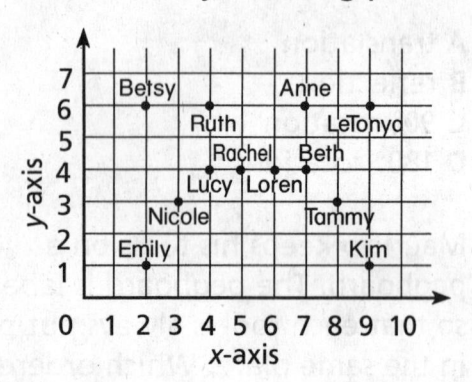

   A (2,6)
   B (3,8)
   C (7,4)
   D (8,3)

8. Jamie was looking at his dresser and noticed that the edges of the drawers formed two types of lines. Which type of lines do the top and bottom edges of the drawer model?

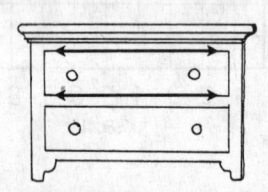

   A intersecting
   B rotated
   C perpendicular
   D parallel

9. Eileen is choosing pattern blocks to complete a puzzle. Two of the blocks must be congruent. Which two blocks appear to be congruent?

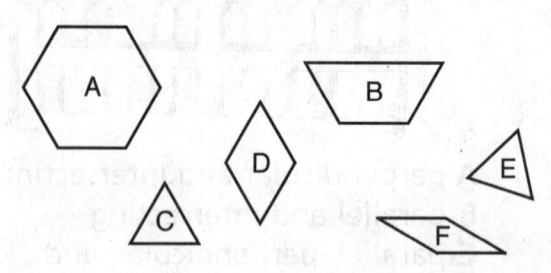

   A A and B
   B B and C
   C C and E
   D D and F

10. The large square shows the quilt pattern called Seesaw Spin. To make the pattern, which transformations were used to move the triangle and then the small square?

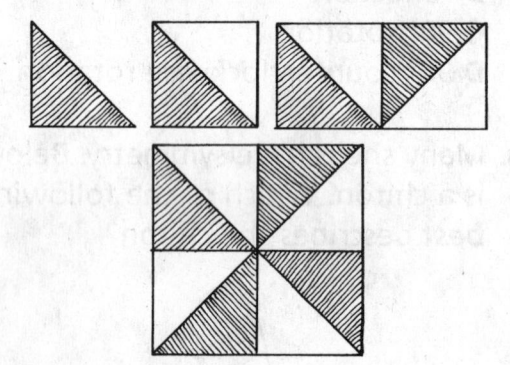

    A translation, 90° rotation
    B reflection, 90° rotation
    C translation, reflection
    D reflection, translation

Go to next page ▶

74 Practice for the EOG Test   Geometry Practice Test

Name _____

GEOMETRY

11. The cables that hold up the Golden Gate Bridge represent which type of lines?

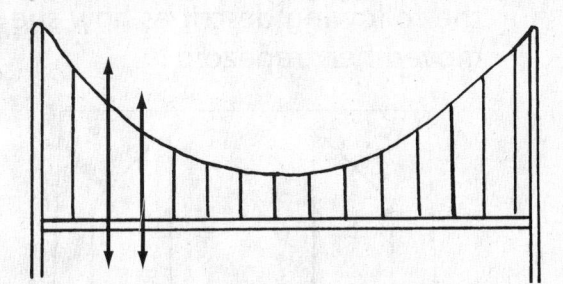

A intersecting   C rotated
B perpendicular D parallel

12. John lives in Hanson Park City. What is located at (7,4)?

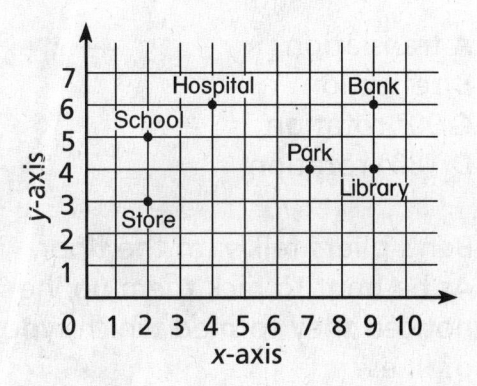

A bank
B library
C park
D store

13. Marge painted a floor cloth using this quilt pattern called Acorns. Which transformation does this pattern model?

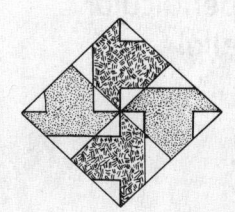

A translation   C 90° rotation
B reflection   D 180° rotation

14. A manual for a stove has a diagram that shows in which direction to turn the knobs. Which of the following describes this transformation?

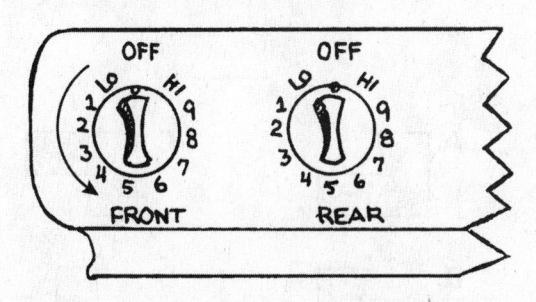

A counterclockwise rotation
B clockwise rotation
C counterclockwise reflection
D clockwise reflection

15. Sally made four shapes on dot paper. Which two shapes are congruent?

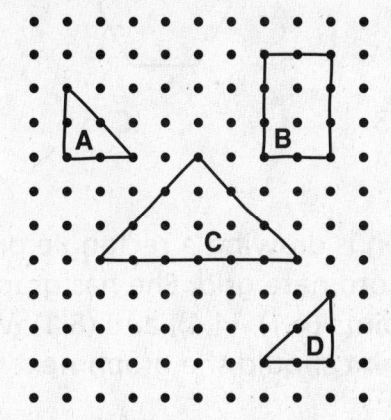

A A and C
B B and C
C A and D
D C and D

Go to next page

Geometry Practice Test   Practice for the EOG Test 75

Name _____

GEOMETRY

16. Jared drew this figure on the board. Which of the following shows how his figure will look after it is translated?

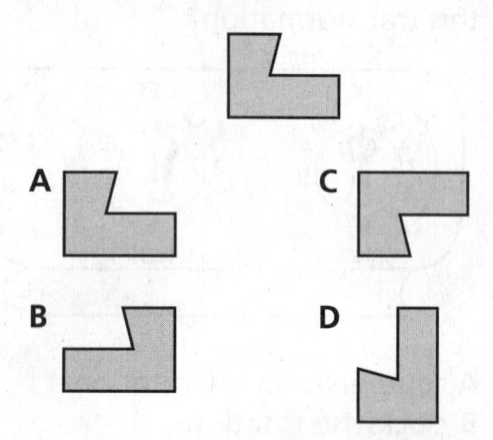

A
B
C
D

17. The honeybee is the state insect of North Carolina. A honeycomb is made of cells in the shape of hexagons. How many lines of symmetry does a hexagon have?

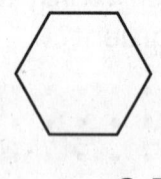

A 3
B 4
C 5
D 6

18. Ren is drawing a rectangle on the coordinate grid. She has graphed points (4,1), (4,4), and (8,1). Which point should she graph next?

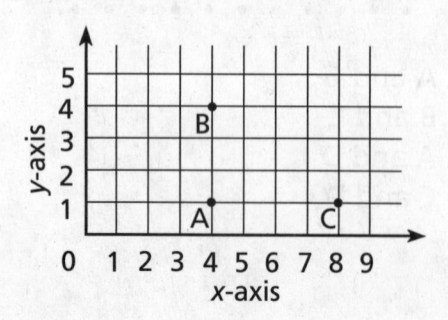

A (6,4)
B (8,4)
C (8,5)
D (4,8)

19. Joan made a trapezoid on a geoboard. Then she moved it from position A to position B. Which of the following describes how she moved her trapezoid?

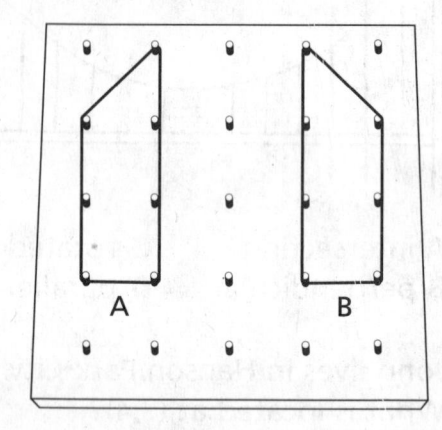

A translation
B reflection
C 90° rotation
D 180° rotation

20. Ben's pliers fell onto the floor. As he bent to pick them up, he noticed they formed which type of lines?

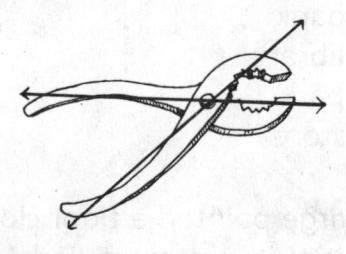

A intersecting
B symmetric
C perpendicular
D parallel

76 Practice for the EOG Test

Geometry Practice Test

Name _____

DATA ANALYSIS & PROBABILITY

Vocabulary

Match the definitions to the correct terms. Record the numbers in the magic square. To check your work, make sure the sums for the columns, rows, and diagonals are the same.

A ___	B ___	C ___
D ___	E ___	F ___
G ___	H ___	I ___

A. the distance between two numbers on the scale of a graph

B. the difference between the greatest and the least number in a set of data

C. a method of gathering information to record data

D. on a graph, areas where the data increase, decrease, or stay the same over time

E. the number(s) or item(s) that occurs most often in a set of data

F. a running total of items being counted

G. the middle number in an ordered set of data

H. the number of times a response occurs

I. a series of numbers placed at fixed distances on a graph to help label the graph

1. cumulative frequency
2. interval
3. frequency
4. median
5. mode
6. survey
7. range
8. scale
9. trends

What is the sum for each row, column, and diagonal?

_____

Data Analysis & Probability Vocabulary Practice   Practice for the EOG Test   77

Name _____

DATA ANALYSIS & PROBABILITY

Vocabulary

Complete the chart. Describe how the terms in each pair are alike and how they are different.

Terms	Alike	Different
bar graph and double-bar graph		
line graph and line plot		
bar graph and circle graph		

Decide whether the terms in each pair are the same, opposite, or not related. Write each pair of terms in the correct column of the chart.

certain—impossible	outcome—survey
likely—unlikely	probability—chance
fair—equally likely	event—likely

Same	Opposite	Not Related

78 Practice for the EOG Test — Data Analysis & Probability Vocabulary Practice

Name _____

DATA ANALYSIS & PROBABILITY

1. Max took a survey of his class. He asked all the students which radio station they liked best. He found 6 students liked WBOP 94 best, 3 students liked WKOL 93 best, and 9 students liked WBAM 92.7 best. Record his survey results in the tally table. How many *more* students liked WBAM 92.7 than WKOL 93?

   _____

   _____
   _____

2. Trina made a Venn diagram to show the relationships among some numbers. What label *best* describes the numbers in each section?

   A   B   C
   (2, 4, 8, 10) (6, 12, 18) (3, 9, 15)

   _____
   _____
   _____
   _____
   _____

3. The Sporting Warehouse sells 4 different colors of T-shirts. How many *more* red and green T-shirts than yellow T-shirts have been stocked?

   T-Shirts Stocked

Blue	👕👕👕
Red	👕👕👕
Green	👕👕👕👕
Yellow	👕👕👕👕👕

   Key: Each 👕 = 4 T-shirts.

   TIP How many T-shirts does each symbol represent?

   _____
   _____

4. **Explain It** Leona made a line graph to show the number of pages she read in her book each night. On which days did Leona read the *most* pages? What conclusion can you draw? Explain your answer.

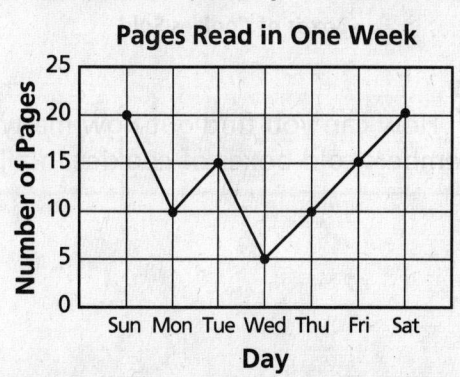

   Pages Read in One Week

   (line graph: Sun=20, Mon=10, Tue=15, Wed=5, Thu=10, Fri=15, Sat=20)

   _____
   _____
   _____
   _____

4.01 Collect, organize, analyze, and display data (including line graphs and bar graphs) to solve problems.

Data Analysis & Probability Practice   Practice for the EOG Test 79

Name _____

DATA ANALYSIS & PROBABILITY

5. Mrs. Wu made a line plot that shows the number of web sites each student used for a research project. How many students used only 3 web sites?

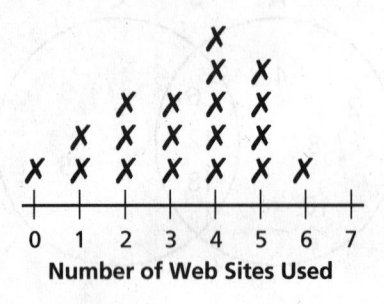

```
                    X
                X X X
            X X X X
        X X X X X X
    X X X X X X X
    +--+--+--+--+--+--+--+
    0  1  2  3  4  5  6  7
       Number of Web Sites Used
```

_____

6. Fernando made this table to show how much money he makes taking care of pets each month. If you made a line graph of the data in Fernando's table, what would the *most* reasonable interval be? Explain your choice.

Fernando's Pet-Sitting Business

Month	Money Earned
May	$29
June	$40
July	$80
August	$70

_____
_____

7. The chess club sold boxes of cookies to raise money one weekend. What fraction of the club members sold 5 boxes each?

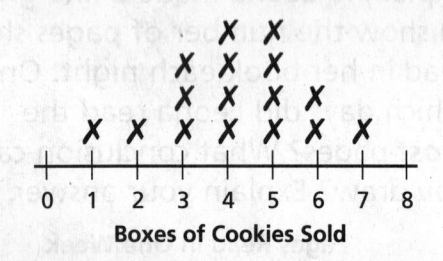

```
                X X
            X X X
        X X X X X
    X X X X X X X X
    +--+--+--+--+--+--+--+--+
    0  1  2  3  4  5  6  7  8
        Boxes of Cookies Sold
```

TIP How can you find out how many members sold boxes of cookies in all?

_____

8. **Explain It** Marta recorded the number of minutes she read each night for five days. She displayed the data in two different ways. Which display is better for finding the number of minutes Marta read on Friday? Explain.

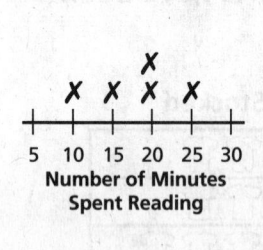

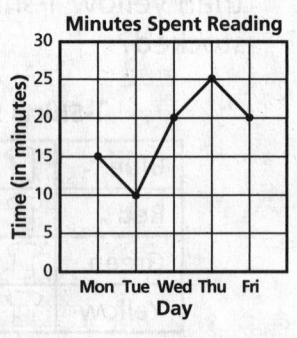

```
                X
        X X X X
    +--+--+--+--+--+--+
    5 10 15 20 25 30
     Number of Minutes
       Spent Reading
```

Minutes Spent Reading
(line graph: Mon 15, Tue 10, Wed 20, Thu 25, Fri 20; Time in minutes 0-30)

_____
_____
_____

4.01 Collect, organize, analyze, and display data (including line graphs and bar graphs) to solve problems.

80 Practice for the EOG Test   Data Analysis & Probability Practice

Name _____

DATA ANALYSIS & PROBABILITY

9. Ruben made a bar graph that shows the life span of some sea animals. What is the life span of the channel bass, North Carolina's state fish?

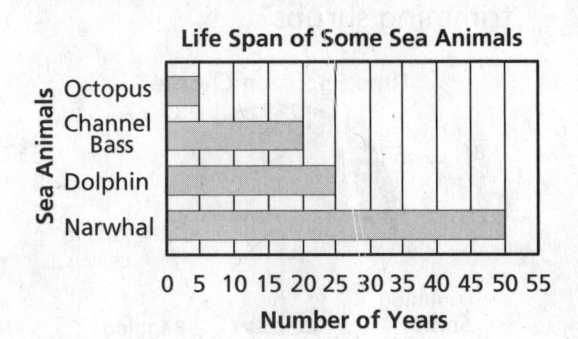

Life Span of Some Sea Animals

(bar graph showing Octopus, Channel Bass, Dolphin, Narwhal on y-axis; Number of Years 0-55 on x-axis)

_____

10. Mrs. Meyer's class voted for their favorite lunchtime drink. The tally table shows the data. Mrs. Meyer uses this data to make a pictograph. How many symbols should she put next to water?

Favorite Drinks

Drink	Tally
Milk	|||| ||||
Juice	||||
Water	|||| ||

Favorite Drinks

Milk	⌐⌐⌐⌐
Juice	⌐⌐
Water	

Key: Each ⌐ = 2 votes.

_____

11. Mr. Roberts asked four students how long it took them to get to school each day. Make a bar graph to compare the number of minutes it takes each student to get to school. Which student takes twice as long as Tom?

Time Traveled to School

Student	Tom	Sam	Ann	Tia
Time	15	25	30	20

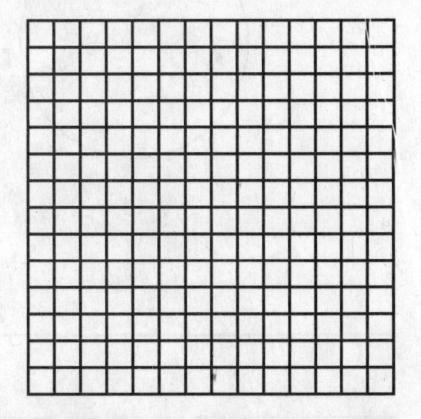

12. **Explain It** Choose which display Kendra should use to compare the heights of some lighthouses in North Carolina. Explain your answer.

TIP What are the differences in the types of information shown in a line plot, a line graph, a bar graph, and a circle graph?

_____
_____
_____
_____

4.01 Collect, organize, analyze, and display data (including line graphs and bar graphs) to solve problems.

Data Analysis & Probability Practice   Practice for the EOG Test   **81**

Name _____

DATA ANALYSIS & PROBABILITY

13. If the Favorite Subjects circle graph shows that 4 students chose math, what is the total number of students surveyed?

    Favorite Subjects
    (circle graph with sections: Reading, Math, Science, Social Studies)

    _____

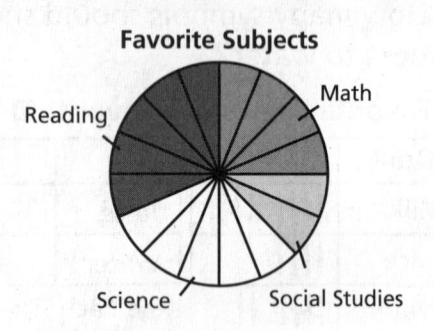

14. Shaheed does chores every Saturday. He made a circle graph to show the kinds of chores he does. What fraction of his time was spent trimming shrubs?

    Time Spent on Chores
    (circle graph: $\frac{4}{8}$ Mowing Lawn, $\frac{1}{8}$ Raking Leaves, $\frac{1}{8}$ Bagging Leaves, ? Trimming Shrubs)

    _____

15. Becky's school sold 4 different types of food at the school carnival. Nachos made up $\frac{1}{10}$ of the sales. Nachos and what other type of food made up $\frac{1}{2}$ of the sales.

    School Carnival Food Sales
    (circle graph with sections: Pizza, Nachos, Hot Dogs, Hamburgers)

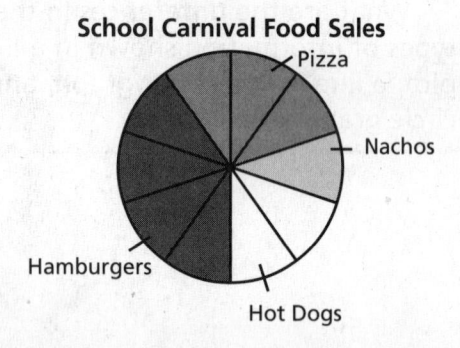

    TIP What is the value of each section of the circle graph?

    _____

16. **Explain It** Seth invited 12 friends to his party. Six of them said they could attend. Four of them said they could not attend. The others have not yet responded. Complete the circle graph to show Seth's data. Explain how you found the number of friends who have not responded.

    Friends Invited to My Party
    (blank circle)

    _____
    _____
    _____

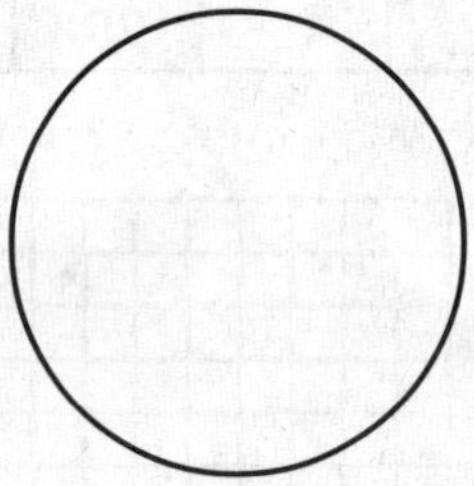

4.01 Collect, organize, analyze, and display data (including line graphs and bar graphs) to solve problems.

82 Practice for the EOG Test

Data Analysis & Probability Practice

Name _____

DATA ANALYSIS & PROBABILITY

17. The U.S. Department of Agriculture made a line graph to show how much wheat has been grown in the United States each year.

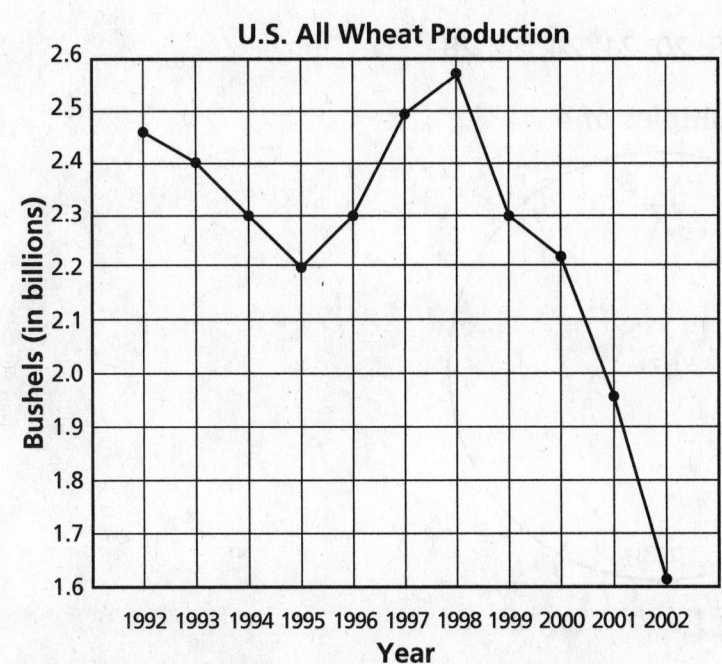

U.S. All Wheat Production

[Line graph showing Bushels (in billions) on y-axis from 1.6 to 2.6, and Year on x-axis from 1992 to 2002. Data points approximately: 1992: 2.46, 1993: 2.40, 1994: 2.30, 1995: 2.20, 1996: 2.30, 1997: 2.49, 1998: 2.57, 1999: 2.30, 2000: 2.22, 2001: 1.96, 2002: 1.62]

Source: USDA-NASS

TIP: How can the direction of the line segments help you draw a conclusion?

a. The line graph shows the production of wheat over time. What time period is displayed on this graph? How many years is this in all?

_____

b. In which year did the United States produce 2.4 billion bushels of wheat? Is this *more than* or *less than* the year before?

_____

c. For which time period did wheat production increase?

_____

d. If the trend continues, what conclusion can you make about wheat production for 2003? Explain.

_____

_____

	PS	Comp	Comm
IV			
III			
II			
I			

4.01 Collect, organize, analyze, and display data (including line graphs and bar graphs) to solve problems.

Data Analysis & Probability Practice   Practice for the EOG Test   **83**

Name _____

DATA ANALYSIS & PROBABILITY

18. Nick wants to make a Venn diagram to show the relationship among these numbers.

    3, 4, 6, 8, 9, 12, 15, 16, 18, 20, 21, 24, 27, 28

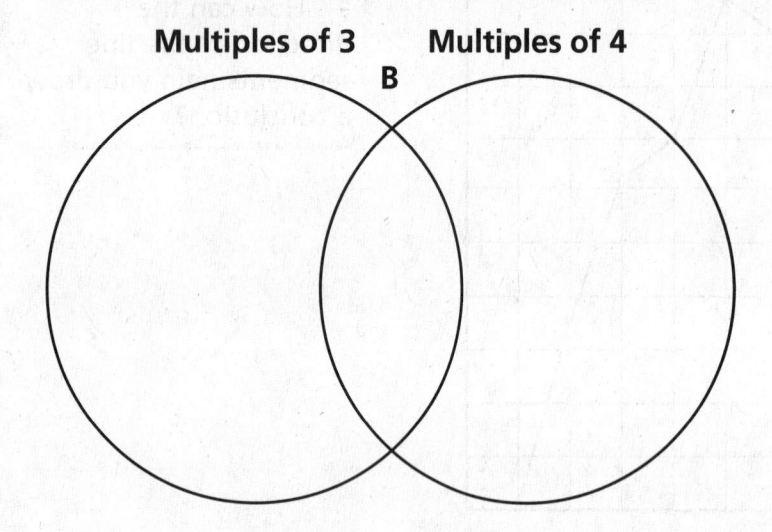

    Multiples of 3   Multiples of 4
                  B

    a. Complete Nick's Venn diagram. Sort the numbers he listed above.

    b. What label could you use to describe the numbers in section B? What numbers are in section B?

    _____

    TIP What does the overlapping part of the two circles tell you?

    c. If you continue to sort numbers, name 3 different numbers you could put in the section labeled Multiples of 3. Name 3 different numbers you could put in section B. Name 3 different numbers you could put in the section labeled Multiples of 4.

    _____

	PS	Comp	Comm
IV			
III			
II			
I			

4.01 Collect, organize, analyze, and display data (including line graphs and bar graphs) to solve problems.

84 Practice for the EOG Test   Data Analysis & Probability Practice

Name _____

DATA ANALYSIS & PROBABILITY

1. Mrs. Suen made a line plot that shows the number of absences each student had in one grading period. What is the mode of this data?

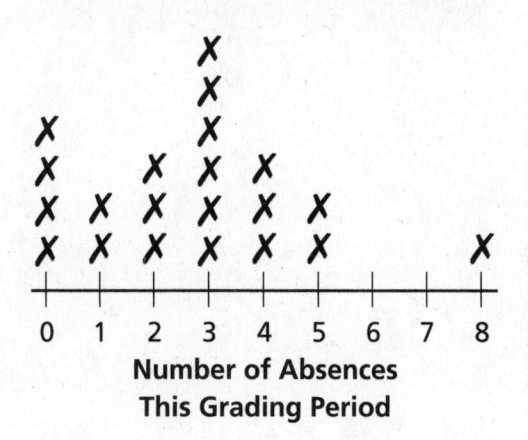

```
            X
            X
            X
    X       X
    X   X   X   X
    X   X   X   X   X
    X   X   X   X   X       X
    +---+---+---+---+---+---+---+---+
    0   1   2   3   4   5   6   7   8
          Number of Absences
          This Grading Period
```

_____

2. Emil found this line graph on the Internet about U.S. corn production. What is the range of this data?

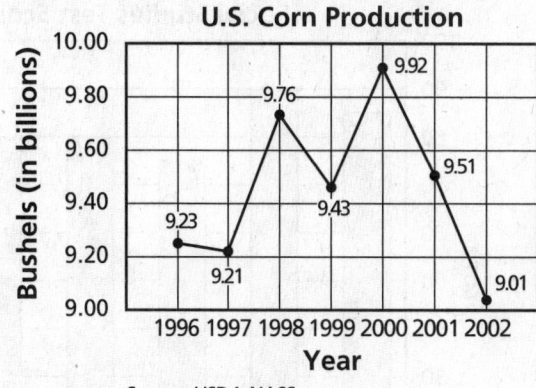

U.S. Corn Production

Bushels (in billions): 9.23, 9.21, 9.76, 9.43, 9.92, 9.51, 9.01
Year: 1996, 1997, 1998, 1999, 2000, 2001, 2002

Source: USDA-NASS

TIP How can reading a line graph help you locate the least and greatest value?

_____

3. Bailey and Raul kept records for one month of the time they spent practicing the piano. How does the median amount of time that Bailey practiced compare with the median amount of time Raul practiced?

Amount Bailey Practiced Each Week

Week	1	2	3	4
Minutes	210	420	225	315

Amount Raul Practiced Each Week

Week	1	2	3	4
Minutes	250	225	360	270

_____

4. **Explain It** Nita kept track of her math grades by writing down her weekly math quiz scores. What is her median grade? Explain how you found the median.

84, 95, 85, 80, 93, 81, 90

_____
_____
_____
_____
_____

4.02 Describe the distribution of data using median, range and mode.

Data Analysis & Probability Practice   Practice for the EOG Test   85

Name _____

DATA ANALYSIS & PROBABILITY

5. Mr. Wilson made a bar graph to show the top six scores his students received on their social studies test about the Tar Heels and their role in the Civil War.

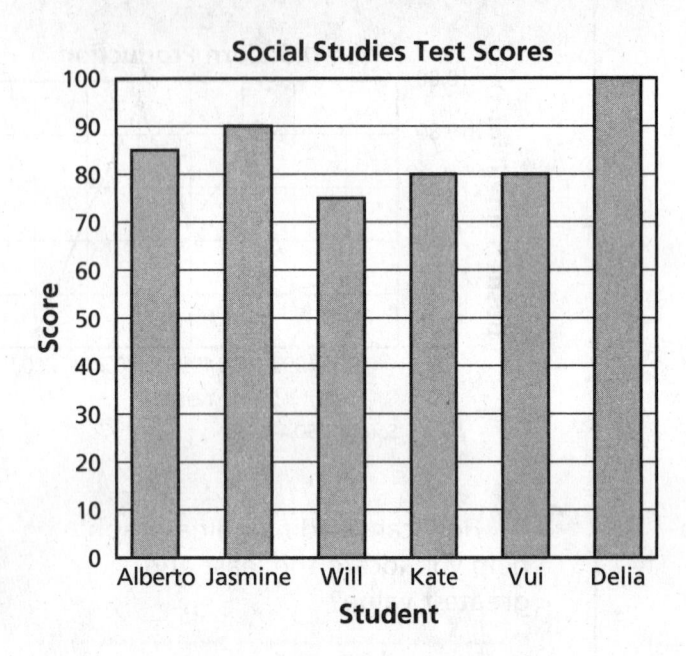

Social Studies Test Scores

(Bar graph showing scores: Alberto 85, Jasmine 90, Will 75, Kate 80, Vui 80, Delia 100)

a. Find the range of the test scores.

_____

b. Find the mode of the test scores.

_____

TIP What is the first step in finding the range on a bar graph?

c. Suppose the range of test scores was 35 points. What score would one student need to receive? Explain.

_____

_____

d. Suppose Kate did not take the test. Which student's score would be the median? What would the median be?

_____

4.02 Describe the distribution of data using median, range and mode.

86 Practice for the EOG Test   Data Analysis & Probability Practice

Name _____

DATA ANALYSIS & PROBABILITY

1. Leo surveyed 19 girls and 19 boys. He made two graphs to show their favorite types of books. How many girls and boys chose biography as their favorite?

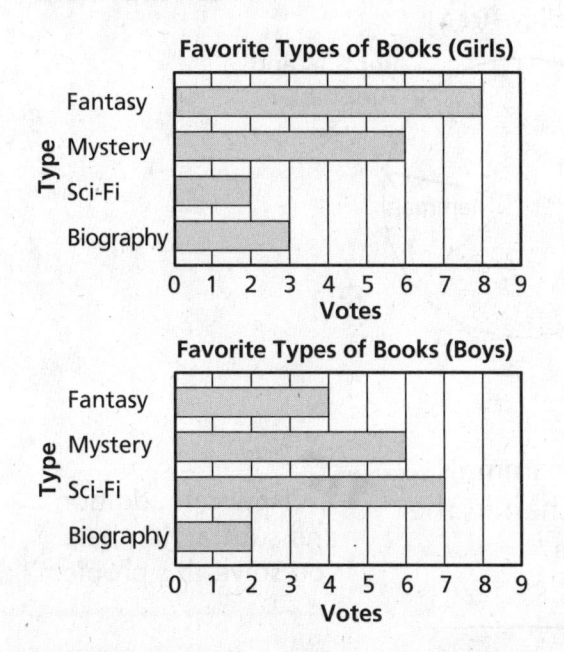

   Favorite Types of Books (Girls)

   Fantasy: 8
   Mystery: 6
   Sci-Fi: 3
   Biography: 2
   (Votes 0–9)

   Favorite Types of Books (Boys)

   Fantasy: 4
   Mystery: 5
   Sci-Fi: 7
   Biography: 2
   (Votes 0–9)

2. For her science project, Beth recorded the daily growth of two plants. Which day did Plant B grow *more* than Plant A? How much more?

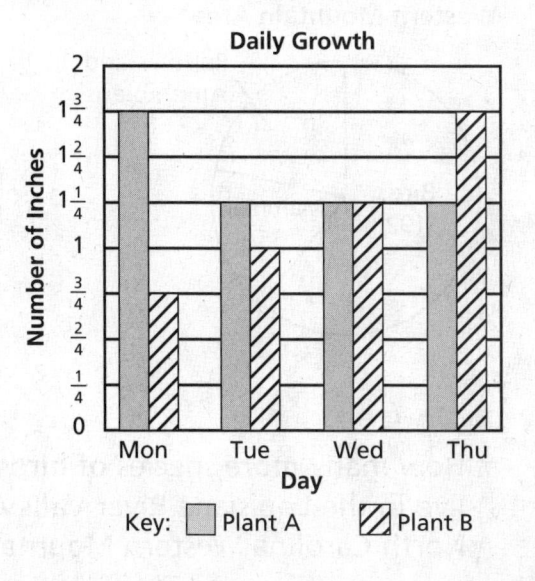

   Daily Growth

   Mon — Plant A: $1\frac{3}{4}$, Plant B: $\frac{3}{4}$
   Tue — Plant A: $1\frac{1}{4}$, Plant B: 1
   Wed — Plant A: $1\frac{1}{4}$, Plant B: $1\frac{1}{4}$
   Thu — Plant A: $1\frac{1}{4}$, Plant B: $1\frac{3}{4}$

   Key: Plant A ▢   Plant B ▨

3. Raphael took a survey at camp to find out the favorite activity for the boys and girls at Camp Laurel.

   Favorite Camp Activities

Activity	Girls	Boys
Canoeing	7	14
Riding	10	7
Hiking	8	12
Swimming	12	10

   For a double-bar graph, choose the *most* reasonable interval for the data in his table—1, 2, or 5? Then find the scale for this interval.

4. **Explain It** Mrs. Stromire recorded the number of fiction and nonfiction books that were checked out each weekday at the library. Which graph makes it easier to compare the number of fiction books and nonfiction books that were checked out on Friday—two bar graphs or a double-bar graph? Explain your answer.

4.03 Solve problems by comparing two sets of related data.

Data Analysis & Probability Practice   **Practice for the EOG Test**

Name _____

DATA ANALYSIS & PROBABILITY

5. These circle graphs show the number of species found living in two different areas of the United States. Study the data.

Species Living in the
North Carolina
Western Mountain Area

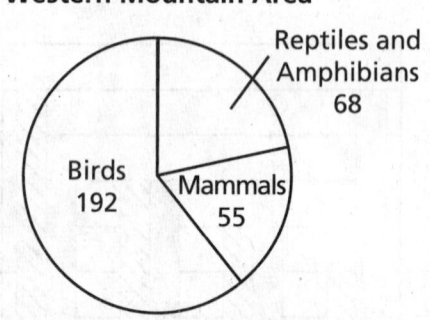

- Reptiles and Amphibians 68
- Birds 192
- Mammals 55

Species Living in the
Louisiana River
Valley Area

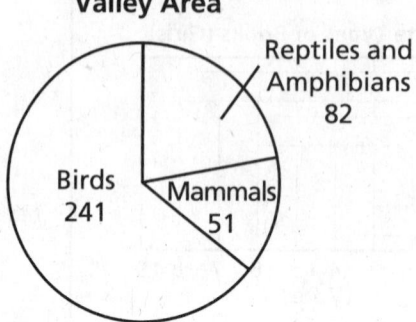

- Reptiles and Amphibians 82
- Birds 241
- Mammals 51

a. How many *more* species of birds and mammals live in the Louisiana River Valley area than in the North Carolina Western Mountain area?

_____

TIP How can addition and subtraction help you solve this problem?

b. Write a problem that can be answered by using the graphs.

_____
_____
_____

c. Find the answer to the problem you wrote. Explain your answer with pictures, words, or numbers.

	PS	Comp	Comm
IV			
III			
II			
I			

4.03 Solve problems by comparing two sets of related data.

88 Practice for the EOG Test   Data Analysis & Probability Practice

Name _____

DATA ANALYSIS & PROBABILITY

1. Derek and Julie are playing a game with a spinner. How many possible outcomes are there if they use the spinner below?

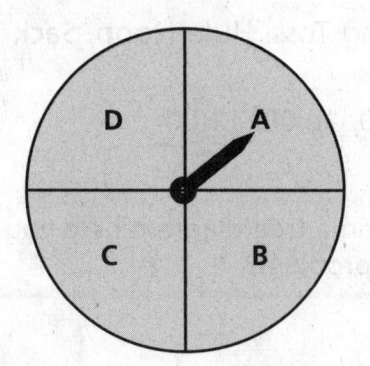

2. Kaitlyn and Mario are playing a game. It is Mario's turn to spin. What is the probability of the pointer stopping on an even number?

TIP How can you write the probability of an event as a fraction?

3. Rita plays a game in which she must spin these two spinners one time each. How many of the possible outcomes include green or blue? Make a list of the possible outcomes.

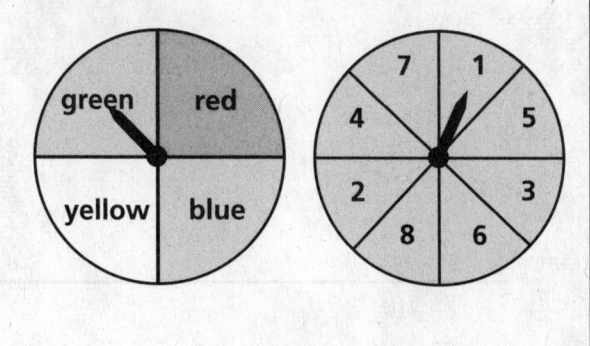

4. **Explain It** Chantelle tosses a number cube labeled 1–6. What is the likelihood that she will toss a 7? Explain.

4.04 Design experiments and list all possible outcomes and probabilities for an event.

Data Analysis & Probability Practice

Practice for the EOG Test 89

Name _____

DATA ANALYSIS & PROBABILITY

5. Mr. Wanguhu keeps a jar of marbles on his desk. If a student picks one without looking, which color marble will he *most likely* pick?

6. Gretel must choose an event and a time from the list below. How many possible combinations does she have?

   Event: Ring Toss, Hula Hoop, Sack Race
   Time: 9:00, 10:00, 11:00

   TIP How can a tree diagram help you solve this problem?

7. Siddarth places these letter tiles in a jar: a, b, c, j, o, p, z. Then he places these number tiles in another jar: 1, 2, 3. He closes his eyes and picks a tile from each jar. How many of the possible outcomes include an odd number and a consonant? Explain.

8. **Explain It** Kiko tosses a coin 10 times. The coin lands heads up 7 times and tails up 3 times. Kiko says the next time he tosses the coin it will *most likely* show heads. Is Kiko correct or incorrect? Explain.

4.04 Design experiments and list all possible outcomes and probabilities for an event.

90 Practice for the EOG Test   Data Analysis & Probability Practice

Name _____

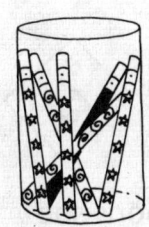

DATA ANALYSIS & PROBABILITY

9. Look at the jars of pencils and erasers below.

   a. Myra closes her eyes and pulls a pencil and an eraser from each jar. If Myra pulls a black pencil and an eraser, draw all the possible outcomes that include a black pencil.

   b. Myra closes her eyes and pulls a pencil from the jar. What is the probability that Myra will pull a pencil with stars?

   _____

   c. Myra closes her eyes and pulls an eraser from the jar. What is the probability that Myra will pull a white eraser?

   _____

   d. Without looking, Myra has a $\frac{1}{6}$ chance of pulling a black pencil from the jar. Which eraser does Myra have a $\frac{1}{6}$ chance of pulling? Explain how you decided.

   _____

   _____

   e. Use the jar at the right. If Myra has a $\frac{1}{3}$ chance of pulling a black marble from the jar, draw what marbles could be in the jar.

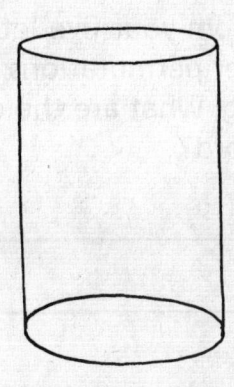

4.04 Design experiments and list all possible outcomes and probabilities for an event.

Data Analysis & Probability Practice   Practice for the EOG Test 91

DATA ANALYSIS & PROBABILITY

Name _____

10. Eiko wrote down on index cards all the possible two-letter permutations she could make from the word *cat*.

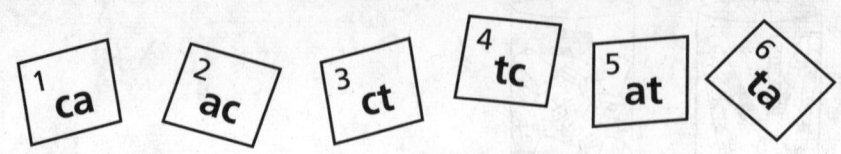

   [1: ca] [2: ac] [3: ct] [4: tc] [5: at] [6: ta]

   a. Suppose Eiko puts all the two-letter permutations she made from *cat* in a jar. Without looking, what chance does she have of pulling a card that is a real word?

   _____

   b. Now she wants to write down on index cards all the possible two-letter permutations she could make using the word *fish*. How many can she make? What are they?

   _____
   _____

   c. Suppose Eiko puts all the two-letter permutations she made from *fish* in another jar. Without looking, what chance does she have of pulling a card that is a real word?

   _____

   d. Now Eiko puts both the *cat* and *fish* permutations together in a jar. What are her chances of pulling one that has the letter *i* on it? Explain.

   _____
   _____

   e. Write a five-letter word. Write all the two-letter permutations from your five-letter word. Suppose you put all the two-letter permutations on index cards in a jar. Without looking, what are the chances of pulling a card that is a real word?

   _____
   _____

4.04 Design experiments and list all possible outcomes and probabilities for an event.

Practice for the EOG Test   Data Analysis & Probability Practice

Name _____

DATA ANALYSIS & PROBABILITY

1. Jolene made a line graph that shows how much money the school store made selling pencils. Between which two weeks did sales drop?

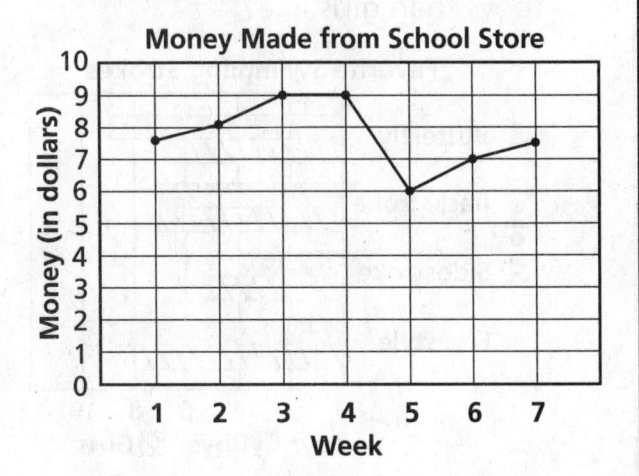

   Money Made from School Store

   A weeks 1 and 2
   B weeks 2 and 3
   C weeks 4 and 5
   D weeks 6 and 7

2. Quincy made a circle graph that shows how he spends his time on weekends. Which activity does Quincy do $\frac{1}{5}$ of the time?

   Weekend Hobbies and Activities

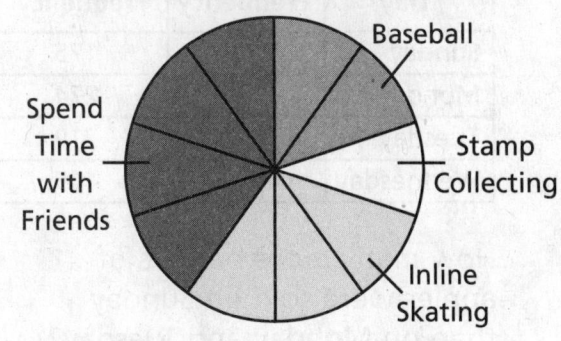

   A Baseball
   B Stamp collecting
   C Inline skating
   D Spending time with friends

3. Which football jersey number is the median number?

   34 45 31 37 58

   A 45   C 34
   B 37   D 31

4. Consuelo made this Venn diagram for the locations of some northern and eastern states. Which of the following is true about the states in Section B?

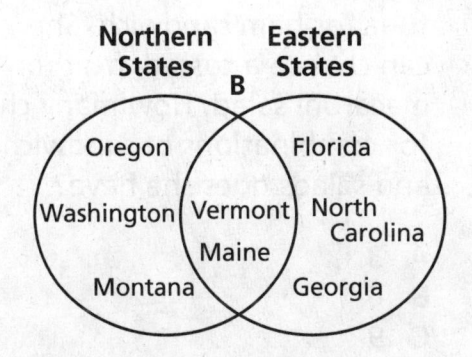

   Northern States / Eastern States
   B
   Oregon, Washington, Montana | Vermont, Maine | Florida, North Carolina, Georgia

   A These states are in the Northwest.
   B These states are in the Northeast.
   C These states are in the Southwest.
   D These sates are in the Southeast.

5. Which kind of graph or plot would be *best* to keep a record of monthly rainfall?

   A line graph
   B circle graph
   C pictograph
   D line plot

Go to next page ▶

Data Analysis & Probability Practice Test   Practice for the EOG Test 93

Name _____

DATA ANALYSIS & PROBABILITY

6. Sandra wrote the numbers 1–10 on cards and placed them in a hat. Without looking, she pulls a card from the hat. Which is the probability of pulling a card with a number less than 3?

   A $\frac{1}{5}$ C $\frac{1}{3}$

   B $\frac{1}{4}$ D $\frac{1}{2}$

7. Noel is deciding which sandwich to order from a menu. Each sandwich comes with a choice of salad. She can choose a turkey, tuna, or ham sandwich. She can choose a tossed, fruit, or macaroni salad. How many choices for combinations of sandwiches and salads does she have?

   A 3
   B 6
   C 9
   D 12

8. This line plot shows how many laps each student in Mrs. Beauchamp's class ran during recess. How many students ran 4 or more laps during recess?

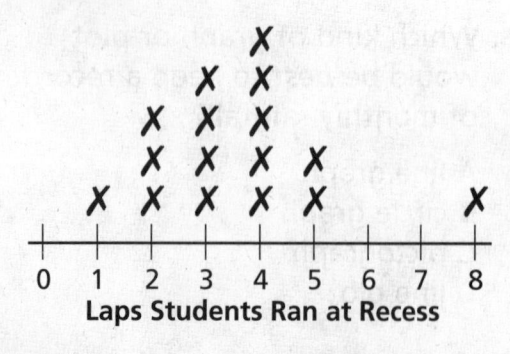

   Laps Students Ran at Recess

   A 1   C 5
   B 3   D 8

9. Harlan took a survey of his class to determine the favorite swimming strokes of fourth graders. He made this double-bar graph to show the results. Which stroke did *fewer* boys than girls like?

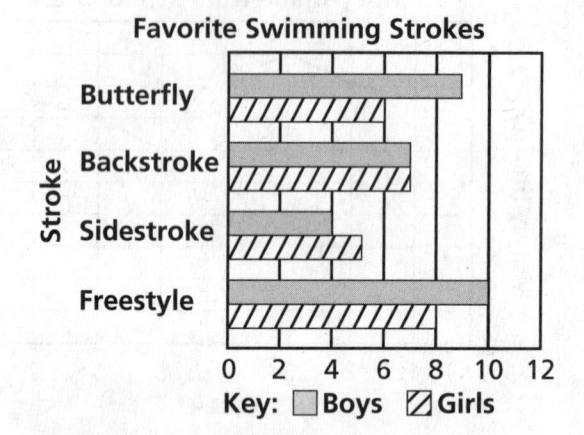

   Favorite Swimming Strokes

   Key: ☐ Boys   ▨ Girls

   A butterfly
   B backstroke
   C sidestroke
   D freestyle

10. The table shows the number of baskets of apples sold each day at Roma's Apple Stand.

    Roma's Apple Stand

Day	Frequency	Cumulative Frequency
Sunday	175	175
Monday	99	274
Tuesday	45	319
Wednesday	126	445

    How many *more* baskets of apples were sold on Sunday than on Monday and Tuesday?

    A 31   C 175
    B 144   D 319

Go to next page ▶

Practice for the EOG Test   Data Analysis & Probability Practice Test

Name _____

DATA ANALYSIS & PROBABILITY

11. If the Birthdays in Our Class circle graph shows that 12 students were born in the spring, how many students were surveyed to collect the data to make the circle graph?

    Birthdays in Our Class

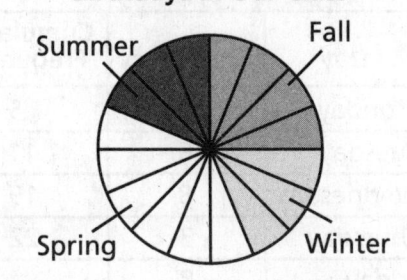

    A 16   C 32
    B 24   D 96

12. Tom organized his data about the average high and low temperatures from May through September in Asheville, North Carolina, in a graph.

    Average Temperatures

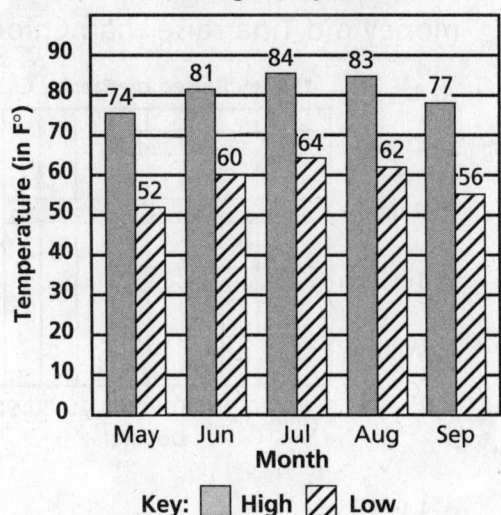

    Key: ■ High ▨ Low

    Which is the difference in the average high and low temperatures in May?

    A 22°F   C 20°F
    B 21°F   D 19°F

13. Morgan is designing a word scramble game. If he uses the word GAME, how many ways can he arrange the letters?

    A 8   C 24
    B 12   D 36

14. Mr. Lighthorse's class counted the total number of fourth graders who were tardy each day for five days. Which number is the mode?

    12, 9, 10, 9, 13

    A 9   C 12
    B 10   D 13

15. The line plots show the number of pets the students in four fourth-grade classes have. Which line plot shows a set of data with a range of 5?

    A Number of Pets—Mr. Smith's Class
    B Number of Pets—Mrs. Reyna's Class
    C Number of Pets—Mr. Watkin's Class
    D Number of Pets—Ms. Patterson's Class

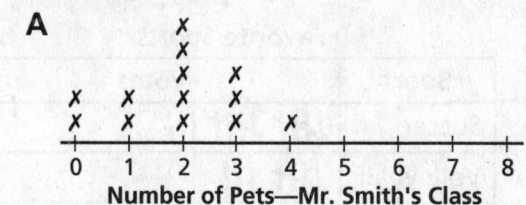

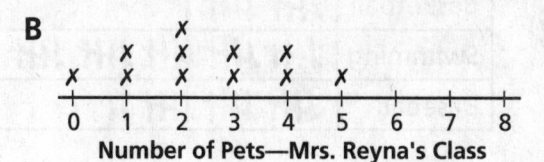

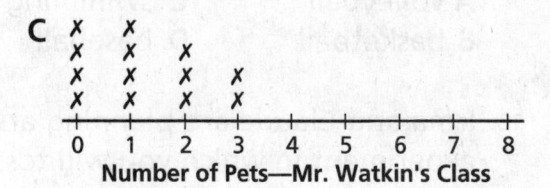

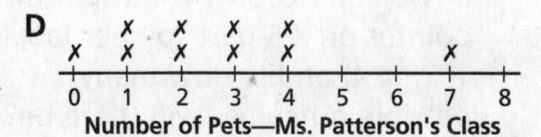

Go to next page ▶

Data Analysis & Probability Practice Test   Practice for the EOG Test   95

DATA ANALYSIS & PROBABILITY

16. Mr. Hill made this table to record the types of fiction books his class read.

    Books Read by Mr. Hill's Class

	Adventure	Mystery	Animal
Boys	12	8	4
Girls	8	11	6

    Which kind of graph or plot would be *best* to compare the books read by the boys in his class with the books read by the girls?

    A double-bar graph
    B line graph
    C circle graph
    D line plot

17. Santos surveyed children in his neighborhood to find their favorite sports. Which sport received twice as many votes as soccer?

    Favorite Sports

Sport	Votes
Soccer	|||| |||| |||
Volleyball	|||| |||
Basketball	|||| ||||	
Swimming	|||| |||| |||| |||| ||||	
Baseball	|||| |||| |||| ||

    A volleyball   C swimming
    B basketball   D baseball

18. Irma and Claude are planning an experiment in which you will toss a two-color counter and spin the pointer on a 5-part spinner labeled 1, 2, 3, 4, and 5. How many possible outcomes will there be?

    A 2   C 10
    B 5   D 12

19. Michael counted the number of students who wore sneakers each day at the pool. What will the cumulative frequency be on Friday?

    Students Wearing Sneakers

Day	Frequency	Cumulative Frequency
Monday	5	5
Tuesday	6	11
Wednesday	8	19
Thursday	3	22
Friday	8	■

    A 30
    B 22
    C 19
    D 14

20. The graph shows the money raised by the Art Club for their annual fund-raiser. How much *more* money did Tina raise than Chloe?

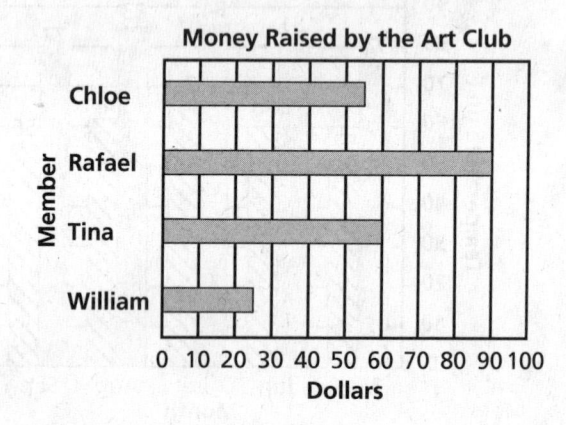

    Money Raised by the Art Club

    A $115
    B $35
    C $30
    D $5

    Go to next page ▶

96 Practice for the EOG Test   Data Analysis & Probability Practice Test

Name _____

DATA ANALYSIS & PROBABILITY

21. Scott recorded the number of tickets he and his friends used at the fair and arranged his data in a graph.

Tickets Used at Fair

Miriam	🎟
Tony	🎟 🎟
Laura	🎟 🎟 🎟 🎟
Paulo	🎟 🎟
Scott	🎟 🎟 🎟

Key: Each 🎟 is 5 tickets.

How many tickets did Laura and Scott use at the fair?

A 15 C 35
B 20 D 60

22. Winona wrote down how many minutes it took her to walk her dog each day for five days. Which is the median number of minutes she walked her dog?

Minutes Walked

Day	1	2	3	4	5
Minutes	22	15	25	18	25

A 15 C 22
B 18 D 25

23. Ian places the letters A, B, C, and D in a jar. He places the numbers 1 and 2 in the same jar. Ian closes his eyes and picks from the jar. What is the probability that Ian will pick a 2?

A $\frac{2}{3}$ C $\frac{1}{4}$
B $\frac{1}{3}$ D $\frac{1}{6}$

24. Kurt's parents own a bike shop. This graph shows the amount of money they made from bikes sold last week. Which statement about the graph is true?

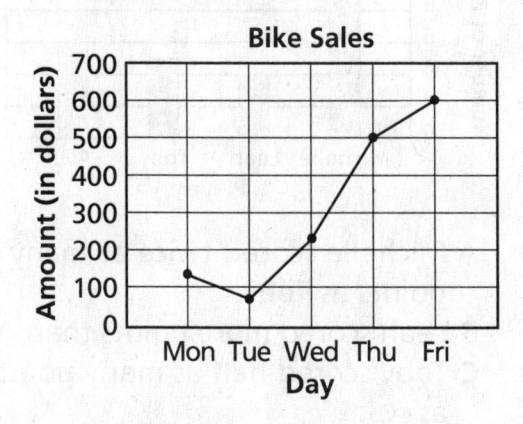

Bike Sales

A They made less than $200 on Wednesday.
B They always made more than $100 each day.
C Bike sales decreased from Tuesday to Wednesday.
D Bike sales increased from Tuesday to Friday.

25. Elena wants to make a circle graph to show how she spends an hour doing homework each day. She spends 15 minutes reviewing math facts and 45 minutes completing other homework. If the circle graph is divided into 4 equal sections, how many sections should Elena shade to show the amount of time she spends reviewing math facts?

A 1
B 2
C 3
D 4

Go to next page ▶

Data Analysis & Probability Practice Test   **Practice for the EOG Test** 97

Name _____

DATA ANALYSIS & PROBABILITY

26. Which statement about the graph is true?

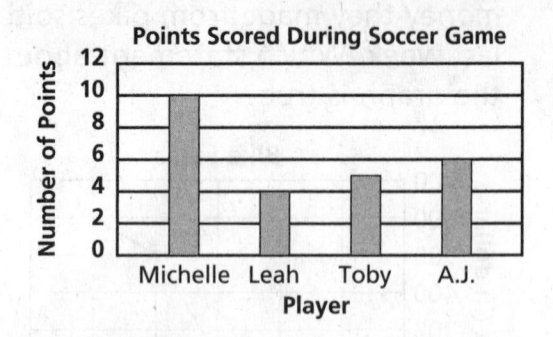

    Points Scored During Soccer Game (bar graph: Michelle 10, Leah 4, Toby 5, A.J. 6)

    A Michelle scored twice as many points as Toby.
    B Leah scored more points than A.J.
    C Toby scored half as many points as A.J.
    D Michelle scored 5 more points than Leah.

27. Darius made these two spinners for a game. How many possible outcomes are there if you spin the pointer on each of his two spinners?

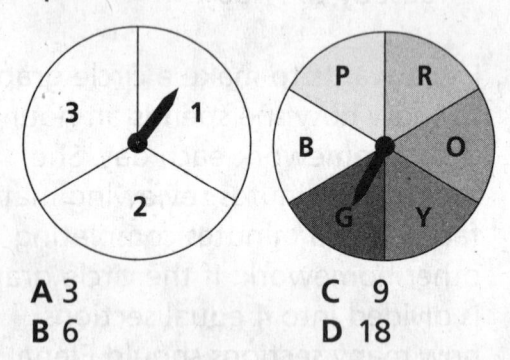

    (Spinner 1: 1, 2, 3; Spinner 2: P, R, O, Y, G, B)

    A 3   C 9
    B 6   D 18

28. Paula is designing a game. How should she label the number cube so that each player will have an *equally likely* chance of tossing an even number or an odd number?

    A 2, 4, 6, 7, 8, 9
    B 2, 3, 4, 5, 6, 7
    C 3, 5, 6, 8, 7, 9
    D 3, 4, 5, 7, 8, 9

29. Two fourth-grade classes voted for a new recess time. How many *more* students in both classes voted for 11:00 than 12:00?

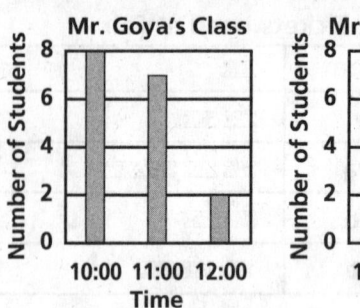

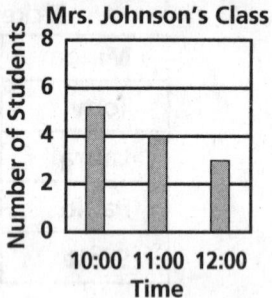

    Mr. Goya's Class (10:00 = 8, 11:00 = 7, 12:00 = 2)
    Mrs. Johnson's Class (10:00 = 5, 11:00 = 4, 12:00 = 3)

    A 3
    B 5
    C 6
    D 11

30. At the school carnival, Mackenzie visited the duck pond booth. Each of the six ducks has the number 1, 2, 3, 4, 5, or 6 written on the bottom of it. No duck has the same number. Mackenzie hopes to pick number 3 or 4 so she can win a special prize. What is the probability that she will pick a 3 or 4?

    A $\frac{1}{2}$
    B $\frac{1}{3}$
    C $\frac{1}{4}$
    D $\frac{1}{5}$

Vocabulary

Write the term for each definition.

Word box: equation, parentheses, expression, variable, inequality

1. a mathematical sentence that shows that two expressions do not represent the same quantity _ _ _ _ _ _ □ _ _

2. a letter or symbol that stands for a number or numbers _ _ □ _ _ _ _ _

3. a part of a number sentence that has numbers and operation signs but does not have an equal sign _ _ _ _ _ □ _ _ _

4. a number sentence that shows that two quantities are equal _ _ □ _ _ _ _ _

Next, rearrange the letters from the boxes above to find the missing term in this sentence.

A □ □ □ □ for the input/output table is: Multiply by 8.

Input	3	4	5	6	7
Output	24	32	40	48	56

Sort the entries in the boxes below into the appropriate column to tell whether each is a variable, an inequality, an expression, or an equation.

x	$5 + 4$	$16 > 11$	$18 + 40 = 58$	n
$25 + 123 < 211$	$33 - a = 15$	h	$81 \div c$	$6 \times 7 = 42$
$22 \times y$	b	$g > 3{,}001$	$45 \div w = 5$	$60 - 15$

Variable	Inequality	Expression	Equation

Algebra Vocabulary Practice

Name _____

ALGEBRA

Vocabulary

Rearrange each set of tiles to show an example of the addition or multiplication property that is named.

1. Commutative Property

   [16] [+] [30] [+] [30] [=] [16]

2. Associative Property

   [(] [6] [)] [+] [9] [=] [6] [+] [(] [5] [)] [+] [9] [+] [5]

3. Identity Property

   [7] [×] [7] [1] [=]

4. Distributive Property

   [20] [20] [×] [+] [13] [(] [)] [10] [3] [=] [×]

Write each expression another way using the given property.

5. 17 + (5 + 4); Associative Property

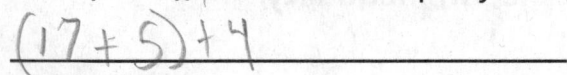

   (17 + 5) + 4

6. 8; Identity Property

   _____

7. 60 × 48; Distributive Property

   _____

8. 67 + 10; Commutative Property

   _____

100 Practice for the EOG Test   Algebra Vocabulary Practice

Name _____

ALGEBRA

1. Marcy divided both sides of this equation by a number and got a new value of 3. What number did Marcy divide by?

   $14 + 7 = 35 - 14$

   _____

2. Steve's little brother Drew has the same amount of money in each hand. He has 2 dimes in one hand and 4 nickels in the other hand. Steve doubles the amount in each hand. What is the value of the money Drew has in each hand now?

   TIP How can you use parentheses to show how each amount increases?

   _____

3. On school field trips, there must be 2 adults for every 8 children. If 12 adults are going on the field trip, how many children are there? Find a rule. Then write the rule as an equation and extend the pattern to find the answer.

Adults	a	2	4	6	8	10	12
Children	c	8	16	24	32	■	■

4. **Explain It** Tim is packing rope into 1-yard coils. If he has 54 feet of rope, how many coils will he end up with? Write an equation that can be used to solve the problem. Explain how you can use the equation to solve the problem.

Feet, f	Yards, y
3	1
6	2
9	3
12	4
15	5

© Harcourt

5.01 Identify, describe, and generalize relationships in which: a) Quantities change proportionally.

Algebra Practice   **Practice for the EOG Test** 101

Name _____

ALGEBRA

5. Webster Township is increasing their fees for trash pickup. Chuck is making a table to show the new fee a customer will pay. Find a rule to describe the change in fees. Write the rule as an equation.

Old Fee	s	$15	$25	$35	$45
New Fee	p	$17	$27	$37	$47

6. Susan saves the same amount of money each week from her checks. If she receives a check for $42, how much money will she have left after she takes out her savings? Use the table to find a rule. Then solve the problem.

Check Amount	g	$42	$48	$50	$52	$54
Amount Left	n	■	$41	$43	$55	$47

TIP How can you test your rule on each pair of numbers in the table?

7. The table shows how much a small bag of popcorn costs at Tradewinds Cinema. If 7 friends go to a movie and each one buys a bag of popcorn, how much will they spend in all? Find a rule. Then write the rule as an equation and find the answer.

Tradewinds Popcorn

Number of Bags	Cost
n	c
2	$6
3	$9
4	$12
5	$15

8. **Explain It** Jessica and Sarah made both sides of the balance equal with 1 apple and 2 plums.

Sarah said that if she adds 1 apple to the left side and 1 plum to the right side, the balance will still be equal. Do you agree or disagree? Explain your answer.

5.01 Identify, describe, and generalize relationships in which: b) Change in one quantity relates to change in a second quantity.

Name _____

ALGEBRA

9. Twin brothers Josh and Joel try to keep the same amount of money in their banks. Right now, they have the coins shown.

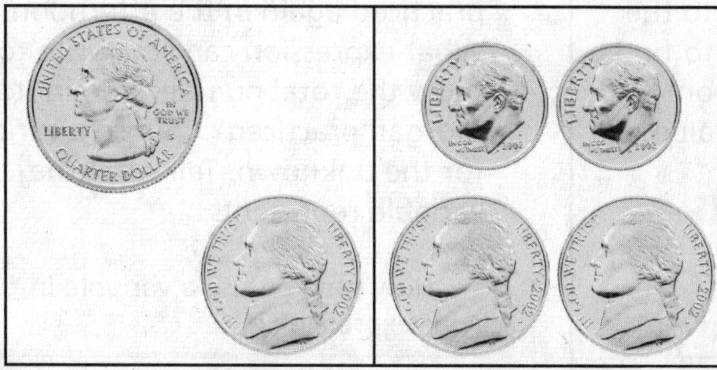

Josh's coins   Joel's coins

a. Write an equation to represent the two amounts.

_____

_____

b. Uncle Harold tripled each amount when he visited on Labor Day weekend. Write an equation to show how much Josh and Joel have now.

TIP How can you use parentheses to help you solve this problem?

_____

_____

c. Joel and Josh each spent half of his money on a pack of markers. Write an equation to show how much each brother spent.

_____

_____

d. Uncle Harold said he would double the amount of money, m, Josh and Joel collect from raking leaves. Josh and Joel wrote this rule and equation to show this.

Multiply by 2. $m \times 2 = t$

Use the rule and the equation to make an input/output table to find the total amount of money, t, they might have.

Input	m				
Output	t				

5.01 Identify, describe, and generalize relationships in which: a) Quantities change proportionally. b) Change in one quantity relates to change in a second quantity.

Algebra Practice   Practice for the EOG Test   103

Name _____

ALGEBRA

1. A toy store has 18 video games on display. It sold 5 of the video games. Then it added 7 new games to the display. Write an expression to find the number of video games on display now. Then find the value of the expression.

_____
_____
_____

2. Morgan practiced the tuba for 35 minutes in the morning. He practiced again in the afternoon. What expression can you write to show the total number of minutes Morgan practiced? Choose a variable for the unknown. Tell what the variable represents.

TIP How can you use a variable in your expression?

_____
_____
_____

3. On Wednesday there were 7 tulips in bloom. By Saturday, there were 21 tulips in bloom. Write an equation to show how many tulips bloomed between Wednesday and Saturday. Choose a variable for the unknown. Tell what the variable represents. Solve the equation.

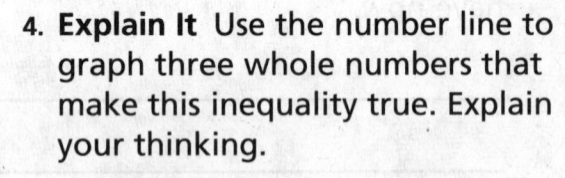

_____
_____
_____
_____

4. **Explain It** Use the number line to graph three whole numbers that make this inequality true. Explain your thinking.

$$n + 4 \geq 16$$

←—+—+—+—+—+—+—+—→
  8 9 10 11 12 13 14

_____
_____
_____
_____

5.02 Translate among symbolic, numeric, verbal, and pictorial representations of number relationships.

104 Practice for the EOG Test   Algebra Practice

Name _____

ALGEBRA

5. Ms. Davenport divided a number of markers, *m*, equally among 5 children. Write an expression to show the total number of markers each child received.

6. Mrs. Penny wrote this expression on the blackboard. Write words to match the expression.

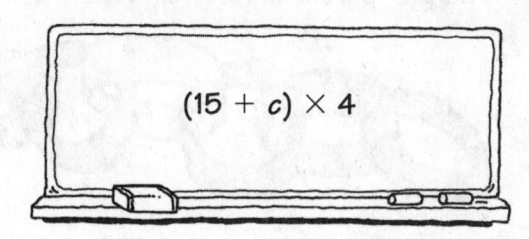

$(15 + c) \times 4$

7. Bennett had 7 trading cards. Then his aunt gave him some packages, *p*, with 5 cards in each package. Bennett now has 42 cards. How many packages did his aunt give him?

Write an equation and work backward to solve.

8. **Explain It** Luis pays $10 for each bag of dog food. Suppose he buys 5 bags of dog food. Luis evaluated this expression to find the total amount he paid for dog food.

$t + 10$
↓
$5 + 10$
↓
15

Explain what error Luis made. Then write the correct answer.

5.02 Translate among symbolic, numeric, verbal, and pictorial representations of number relationships.

Algebra Practice Practice for the EOG Test **105**

Name _____

ALGEBRA

9. The math club has an origami booth at the school fair. Club members are decorating the booth with paper chains.

purple purple purple yellow purple purple purple yellow purple purple purple yellow

___ ___ ___ ___ ___ ___ ___ ___ ___ ___ ___ ___

a. Number each link in the chain beginning with 1. Write the numbers under the links.

b. What number pattern can you see under the yellow links?

_____

c. What color will chain number 16 be?

_____

d. Color the chain to show a pattern that uses multiples. Number each link in the chain beginning with 1. Write the numbers under the links. Describe your pattern. What color would the next four links be?

___ ___ ___ ___ ___ ___ ___ ___ ___ ___ ___ ___

_____

_____

_____

	PS	Comp	Comm
IV			
III			
II			
I			

5.02 Translate among symbolic, numeric, verbal, and pictorial representations of number relationships.

106 Practice for the EOG Test   Algebra Practice

Name _____

ALGEBRA

1. Julio has 1 quarter, 2 dimes, and 1 nickel. Kim has 2 quarters and 1 dime. Do they have the same amount of money? Write equations to explain your answer.

   TIP How much is each coin worth?

2. Alice is taking some money with her to the store, but she wants to leave an equal amount in her bank. What must Alice do to make the values equal?

   Money taken | Money left

3. Mayra spent $21 on some calendars. Each calendar costs $7. Write an equation to find how many calendars she bought. Choose a variable for the unknown. Tell what the variable represents. Use mental computation to solve, and then explain how you can check your answer.

4. **Explain It** Three cousins raced the 200-yard dash. Jake ran it in $\frac{6}{12}$ of a minute, Justin in $\frac{3}{4}$ of a minute, and Jed in $\frac{5}{6}$ of a minute. Graph the fractions on the number line to show who ran the fastest.

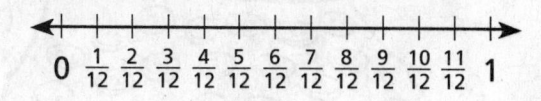

   0 $\frac{1}{12}$ $\frac{2}{12}$ $\frac{3}{12}$ $\frac{4}{12}$ $\frac{5}{12}$ $\frac{6}{12}$ $\frac{7}{12}$ $\frac{8}{12}$ $\frac{9}{12}$ $\frac{10}{12}$ $\frac{11}{12}$ 1

   Explain how you know who ran fastest.

5.03 Verify mathematical relationships using: a) Models, words, and numbers.

Algebra Practice

Name _____

ALGEBRA

5. Rochelle told her little sister that she knew what 1,245 × 1 equals without using paper and pencil. Her sister didn't believe her. Which multiplication property did Rochelle use to solve the problem?

TIP Which property states that the product of 1 and any number is that number?

_____

6. The fourth-grade class arranged cartons of juice into 12 groups of 2, 5 times. Show how you can use the Associative Property to find out how many cartons of juice the class had in all. Find the product.

_____

7. Mrs. Avila is making cookies for five children. There are 16 cookies cooling on paper. She places 4 more cookies on the paper to cool. When the cookies cool, Mrs. Avila divides them into 5 equal groups. Write an expression to find the number of cookies each child gets. Then follow the order of operations to find the value. Write the correct order of operations.

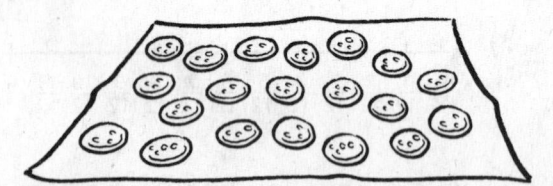

_____

8. **Explain It** Emma has 18 paintings for an art show. She wants to hang them in 3 groups of 6. The wall is not long enough for 6 across, so Mrs. Rosas says to hang the paintings in 6 groups of 3 instead.

Emma says there are not enough paintings for 6 groups of 3. Is Emma correct? Explain by using one of the properties of multiplication.

_____

5.03 Verify mathematical relationships using: b) Order of operations and the identity, commutative, associative, and distributive properties.

108 Practice for the EOG Test

Algebra Practice

Name _____

ALGEBRA

9. Kintha solved this expression using mental computation.

46 + 22 + 54 + 28 = 150

Change the order or group the addends to show how she might have added them mentally. Tell which addition properties she used.

_____
_____
_____

10. Julian and Kyle are planting trees over a large area of land. The diagram shows how many rows of two kinds of trees are being planted. Use the Distributive Property to find how many trees will be planted in all.

            30   4
        9

_____
_____

11. The Growing Fields Nursery sold an average of 72 shrubs each day for 7 days. This is 40 more shrubs than last year during the same 7 days. How many shrubs did they sell last year during the same 7 days?

TIP How can you use the Distributive Property to solve this problem?

12. **Explain It** Shannon took a calculator to the store with her. She keyed in the following: $7 + $11 × 2. The calculator displayed $36. Is the calculator correct? Explain.

7+11×2

_____
_____
_____

5.03 Verify mathematical relationships using: b) Order of operations and the identity, commutative, associative, and distributive properties.

Algebra Practice   Practice for the EOG Test 109

Name _____

ALGEBRA

13. Miguel has boxes of tiles with 4, 6, 9, 10, 12, and 16 tiles.

   a. Which of the boxes of tiles can be used to make a square if all the tiles are used? Draw square arrays to explain your answer.

   _____

   b. Which two boxes of tiles combined can be used to make a square if all the tiles are used? Draw a square array to explain your answer. How many tiles are in your square array?

   _____
   _____

   c. Which three boxes of tiles combined can be used to make a square if all the tiles are used? Draw a square array to explain your answer. How many tiles are in your square array?

   _____
   _____

5.03 Verify mathematical relationships using: a) Models, words, and numbers.

110 **Practice for the EOG Test**

Algebra Practice

Name _____

ALGEBRA

14. The marching band at Eastview High is trying to design a new arrangement for their group. Right now, they have the majorettes in front of the musicians.

   a. How many band members are there in all? Use the Distributive Property to find the product.

   [Diagram: a 4×20 block of X's labeled 20 on top, and a 4×5 block of X's labeled 5 on top and 4 on the side]

   _____

   b. If the band rearranges the majorettes as shown below, will the total number of band members change? Use the Distributive Property to explain your answer.

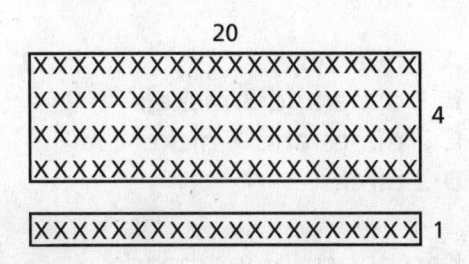

   [Diagram: a 4×20 block of X's labeled 20 on top and 4 on the side, and a 1×20 block of X's labeled 1 on the side]

   _____

   c. The band decided to put the majorettes in the middle as shown at the right. Will the total number of band members change? Use the Distributive Property to explain your answer.

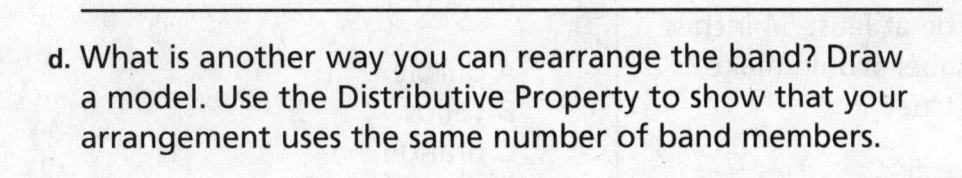

   [Diagram on right: a 4×10 block of X's labeled 10 on top and 4 on the side, a 2×10 block labeled 2 on the side, and a 4×10 block labeled 4 on the side]

   _____

   d. What is another way you can rearrange the band? Draw a model. Use the Distributive Property to show that your arrangement uses the same number of band members.

	PS	Comp	Comm
IV			
III			
II			
I			

5.03 Verify mathematical relationships using: b) Order of operations and the identity, commutative, associative, and distributive properties.

Algebra Practice Practice for the EOG Test **111**

Name _____

ALGEBRA

1. Mallori needs to balance the scale to show a true equation.

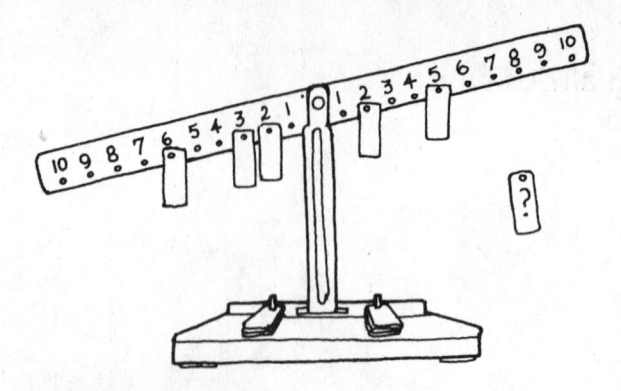

   Which weight can she add to the right side to balance the scale?

   A 2
   B 3
   C 4
   D 6

2. A color printer makes 6 copies, c, per minute, m. Which is a rule for the pattern of printed copies shown in the table?

Input	m	1	5	10	15	20
Output	c	6	30	60	90	120

   A $6 \times m = c$
   B $6 \times c = m$
   C $c + 6 = m$
   D $m + 6 = c$

3. To ride Mantis, a roller coaster at Cedarpoint Amusement Park, a person must be at least 54 inches tall. Which number would make this inequality true?

   $h \geq 54$

   A 45
   B 50
   C 53
   D 61

4. Angie wants to put 32 pairs of shoes into a shoe rack with 30 single slots. Which expression can be used to find how many shoes will be left over?

   A $2 \times 32 - 30$
   B $2 \times (32 - 30)$
   C $2 \times 32 + 30$
   D $2 \times (32 + 30)$

5. Prem has 1 dime and 1 quarter in his left pocket. He has 1 half dollar and 1 nickel in his right pocket. Which coins can Prem add to his left pocket so that the coins in both pockets have the same value?

   A 1 quarter
   B 1 dime and 3 nickels
   C 2 dimes and 1 nickel
   D 2 dimes

6. Karen is stringing beads.

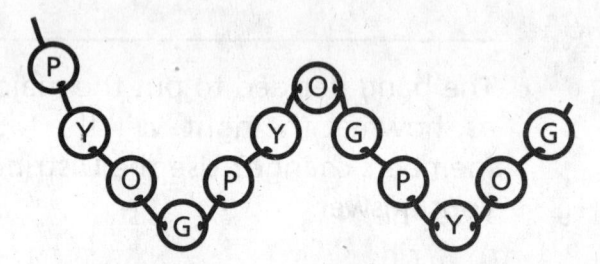

   Which color will the sixteenth bead be?

   A purple
   B yellow
   C orange
   D green

Go to next page ▶

112   Practice for the EOG Test   Algebra Practice Test

Name _____

ALGEBRA

7. Louis wrote two different equations to show how much two rocks in his garden weigh altogether. He says that he can use either equation to find the total weight. Which addition property did he use?

$$n = 85 + 50 \qquad n = 50 + 85$$

A Associative Property
B Zero Property
C Identity Property
D Commutative Property

8. The San Diego Zoo has a Komodo dragon that is about 6 feet long. Which equation can be used to find about how many inches long the Komodo dragon is? Let f represent the number of feet and i represent the number of inches.

A $f \div 12 = i$
B $f + 12 = i$
C $f \times 12 = i$
D $f - 12 = i$

9. David is helping his little sister with her homework. He writes the following expression on a sheet of paper. Which statement could the expression match?

$$83 - (14 \times 4)$$

A eighty-three pens plus fourteen packages of four pens
B eighty-three pens minus fourteen packages of four pens
C four packages of fourteen pens plus eighty-three pens
D four packages of fourteen pens minus eighty-three pens

10. Look at these equations.

$$\blacklozenge + \blacklozenge = 8$$
$$\heartsuit \times \heartsuit = 9$$

Which is the product of $\blacklozenge$ and $\heartsuit$?

A 7
B 12
C 17
D 72

11. Alana is buying a calculator. She needs one that follows the order of operations. She uses some expressions to test each calculator. Based on the following results, which one should she buy?

A $32 + 6 \times 3$ and the display shows 114.
B $24 \div 6 - 2$ and the display shows 6.
C $58 + 6 \div 2$ and the display shows 61.
D $30 - 5 \div 5$ and the display shows 5.

12. Mark wrote this equation to show one of the properties of multiplication. Which property does the equation show?

$$(2 \times 6) \times 4 = 2 \times (6 \times 4)$$

A Associative Property
B Zero Property
C Commutative Property
D Identity Property

Go to next page ▶

Algebra Practice Test

Name _____

ALGEBRA

13. A ticket service adds a $3 fee to every ticket purchased. The table shows the cost of tickets before and after the fee. Which is a possible rule for the pattern shown in the table?

Ticket Price	t	$18	$36	$54	$72
Total Cost	c	$21	$39	$57	$75

A $c = t + \$3$
B $t = c + \$3$
C $c = t \times \$3$
D $t = c \times \$3$

14. A wildlife park restocks the fish in its ponds. Workers put 48 fish in each of 5 ponds. Which operations can be used in the circles below to show how many fish were used in all?

5 ● 48 = (5 × 40) ● (5 × 8)

A × and ×
B + and −
C × and +
D + and ÷

15. Adam has 24 magnets in the shape of states. Each magnet costs $2. Which expression could be used to find the total amount Adam paid for them?

A 24 + $2
B 24 − $2
C 24 ÷ $2
D 24 × $2

16. Tarran bought 2 smoothies and a carrot muffin. Which expression could be used to show the total amount of money Tarran spent?

Wright Street Snacks

Snack	Cost
Smoothies	$4
Pretzel	$2
Carrot Muffin	$3

A 4 × 2 × 3
B 2 × 4 + 3
C 4 ÷ 2 + 3
D 2 × (4 + 3)

17. Leslie has 13 toy animals. She gives 4 to her little sister and then buys some more at a craft fair. Let n represent the number of toy animals she buys. Which expression shows how many toy animals Leslie has now?

A 13 + 4 + n
B 13 − (4 − n)
C 13 − 4 + n
D 13 + 4 − n

18. Max's Bakery had 58 loaves of bread when it opened. By noon there were only 19 loaves left. Let b equal the number of loaves sold before noon. Which value of b makes this equation true?

58 − b = 19

A 19
B 29
C 39
D 49

Go to next page ▶

114 Practice for the EOG Test

Algebra Practice Test

Algebra Practice Test

19. A baby goat at birth weighs about 45 pounds. Two baby goats grow up to weigh a total of 250 pounds. Let *w* represent the weight the two baby goats gained. Which equation could be used to find out about how much weight they gained in all?

    A $w - (45 \times 2) = 250$
    B $w \times (45 \times 2) = 250$
    C $w + (44 \div 2) = 250$
    D $w + (45 \times 2) = 250$

20. The table shows how a factory divides bottles into equal groups for packaging. Which is a possible rule for the pattern shown in the table?

Total Bottles	*t*	96	88	80	72
Number of Cartons	*n*	12	11	10	9

    A Divide by 8.
    B Multiply by 8.
    C Subtract 8.
    D Add 8.

21. Aaron wants to buy 3 pairs of jeans. Each pair costs $28.50 with tax. Which equation can Aaron use to find what the total cost, *c*, will be?

    A $\$28.50 \div 3 = c$
    B $\$28.50 \times 3 = c$
    C $\$28.50 + \$28.50 = c$
    D $\$85.50 \div 3 = c$

22. The members of a track team ran a total of 3 miles. They used the following table to find out how many yards they ran. Use $y = m \times 1{,}760$ to complete the table. How many yards did the team members run?

    m = number of miles
    y = number of yards

Miles	*m*	1	2	3
Yards	*y*	1,760	■	■

    A 3,520 yards   C 7,040 yards
    B 5,280 yards   D 8,800 yards

23. A snack shop offers a discount between midnight and 3:00 A.M. Which equation can you use to complete the table?

    Red Eye Snack Shop

Price, *p*	Discount Price, *d*
$10	$8
$15	$13
$20	$18
$25	■

    A $p - 2 = d$ C $p + 2 = d$
    B $d \div 2 = p$ D $p \times 2 = d$

24. Sam bragged to his little brother that he knew what $99{,}583 + 0$ is without using a calculator. Which addition property is Sam using?

    A Distributive Property
    B Associative Property
    C Identity Property
    D Commutative Property

Go to next page ▶

Name _____

ALGEBRA

25. Which numbers are missing in this ratio table?

Input	x	2	4	6	■
Output	y	6	12	■	24

A 18 and 8
B 18 and 12
C 14 and 8
D 24 and 8

26. Which of the following shows the Associative Property of Addition?

A $7 + 3 = 7 + 3$
B $7 \times (10 + 2) = (7 \times 10) + (7 \times 2)$
C $(7 + 3) + 5 = 7 + (3 + 5)$
D $0 + 7 = 7$

27. The 72 fourth graders lined up in one row. Then their teacher said, "Find the product of 72 × 1 mentally." Which property should the students use to solve the problem?

A Identity Property
B Zero Property
C Commutative Property
D Associative Property

28. The Number Cruncher begins with a number 144 and an output of 12. The next two numbers to be input are 132 and 120. The next two output numbers are 11 and 10. If 108 is input, which number is the output?

A 12
B 9
C 8
D 7

29. If Sally's mom makes 12 pints of juice, how many cups of juice does she make? Which equation can be used to find the number of cups of juice she makes?

p = pints, c = cups

A $c = p \times 2$
B $c = p + 2$
C $c = p - 2$
D $c = p \div 2$

30. Ibrahim made this model with tiles. He used the same number of tiles for each array.

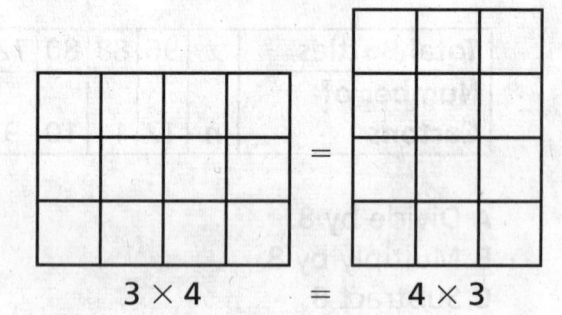

3×4 = 4×3

Which multiplication property did he model?

A Identity Property
B Associative Property
C Commutative Property
D Distributive Property

STOP

116 Practice for the EOG Test

Algebra Practice Test

Name _____

EOG PRACTICE TEST

1. When a number is put into the number machine, a different number comes out.

   If a 20 goes in, a 40 comes out. If a 200 goes in, a 400 comes out. If a 2,000 goes in, what number should come out?

   A 400
   B 2,000
   C 4,000
   D 24,000

2. To the nearest pound, Reba bought 5 pounds of ground meat. Which of the following amounts could be the weight that was marked on the package of meat she bought?

   A 4.2 pounds
   B 4.4 pounds
   C 5.4 pounds
   D 5.6 pounds

3. Brad wants to put a wallpaper border around the walls of his room, just below the ceiling. How many feet of border does he need?

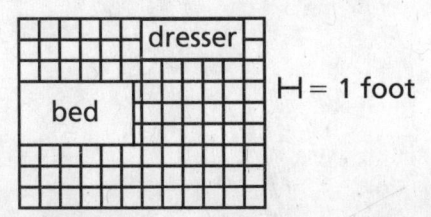

   ☐ = 1 foot

   A 36 feet   C 48 feet
   B 42 feet   D 108 feet

4. Which two streets on the map appear to be perpendicular?

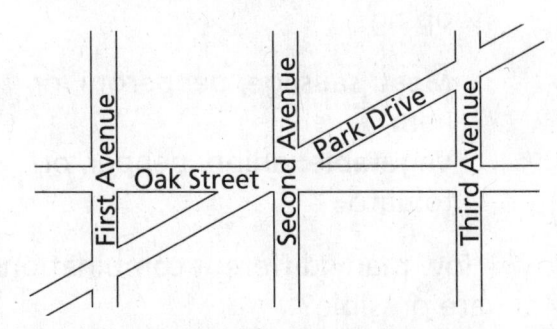

   A First Avenue and Oak Street
   B First Avenue and Second Avenue
   C Third Avenue and Park Drive
   D Park Drive and Oak Street

5. Which is the median number of pennies saved?

   Number of Pennies Saved

Sun	Mon	Tue	Wed	Thu	Fri	Sat
12	15	18	12	14	11	16

   A 11
   B 12
   C 14
   D 18

6. Mari had 35 baseball cards. She gave 14 to her brother and then gave some to her sister. Let c represent the number of cards she gave her sister. Which expression shows how many cards she has left?

   A $35 - 14 + c$
   B $35 + 14 - c$
   C $35 + (14 - c)$
   D $35 - (14 + c)$

Go to next page ▶

Practice for the EOG Test 117

Name _____

EOG PRACTICE TEST

7. Sharon can choose one meat and one vegetable for her pizza topping.

   Meat: sausage, pepperoni, or ham
   Vegetable: onion, pepper, or tomato

   How many different combinations are possible?

   A 3   C 9
   B 6   D 12

8. The coordinate grid shows where some students sit in the classroom. Which ordered pair shows where Pat sits?

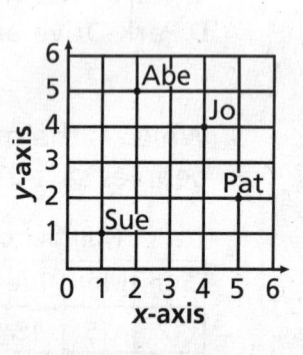

   A (2,5)   C (5,2)
   B (1,1)   D (4,4)

9. James ate 3 brownies from the pan shown below.

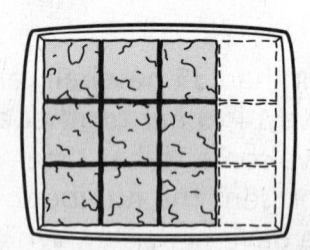

   What fraction of the brownies did he eat?

   A $\frac{1}{4}$ C $\frac{1}{2}$
   B $\frac{1}{3}$ D $\frac{3}{3}$

10. Mrs. Robins used a total of 1 yard of fabric to cover a pillow. She used $\frac{3}{8}$ yard of blue fabric, $\frac{2}{8}$ yard of yellow fabric, and some green fabric. What fraction of a yard of green fabric did she use?

    A $\frac{1}{8}$ yard
    B $\frac{1}{4}$ yard
    C $\frac{3}{8}$ yard
    D $\frac{5}{8}$ yard

11. Allison drew a diagram of her flower garden.

    4 feet
    5 feet

    Her vegetable garden is 2 times as long and 2 times as wide as her flower garden. What is the area of her vegetable garden?

    A 18 square feet
    B 20 square feet
    C 36 square feet
    D 80 square feet

    Go to next page ▶

118 Practice for the EOG Test   EOG Practice Test

Name _____

EOG PRACTICE TEST

12. The table shows the heights of some mountains in North Carolina.

Mountains of North Carolina

Mountain	Height (in feet)
Clingmans Dome	6,643
Grandfather Mountain	5,984
Mount Mitchell	6,684
Stone Mountain	2,305
Wayah Bald	5,385

If you were going to use the data to make a bar graph, which would be the *most* reasonable interval for the scale?

A 1
B 10
C 100
D 1,000

13. The array of cards below is used by football fans for a display during a half-time show. Which square number does the array show?

A 25
B 36
C 49
D 64

14. Marissa collected this data of low temperatures for the first five days in April.

April 2003

Date	Temperature (in °F)
1	32
2	39
3	39
4	41
5	35

Which is the mode for the data Marissa collected?

A 41
B 39
C 35
D 32

15. Which is a possible rule for the pattern shown in this table?

Input	a	10	8	6	4
Output	b	6	4	2	0

A $a + 4 = b$
B $a \times 4 = b$
C $a \div 4 = b$
D $a - 4 = b$

16. Mary keeps her library card in her wallet. Which is the *best* estimate for the length of her library card?

A 8 millimeters
B 8 centimeters
C 8 decimeters
D 8 kilometers

Go to next page ▶

EOG Practice Test   Practice for the EOG Test

Name _____

EOG PRACTICE TEST

17. The cafeteria staff asked 32 students to choose their favorite lunch—hot dogs, pizza, or hamburgers. Which conclusion can you make about the data, as shown in the circle graph?

    Favorite School Lunches

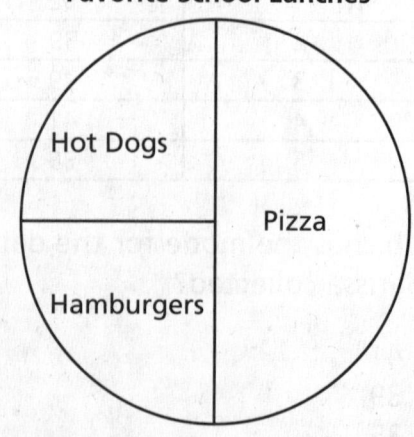

    Hot Dogs
    Pizza
    Hamburgers

    A More students chose hot dogs than hamburgers.
    B More students chose hamburgers than hot dogs or pizza.
    C More students chose pizza than hot dogs.
    D Fewer students choose hot dogs than hamburgers.

18. At 81 yards, Brett Favre has the longest completed pass in Super Bowl history. David Woodley has the third longest completed pass at 76 yards. How much longer was Brett Favre's pass?

    A 10 yards
    B 7 yards
    C 6 yards
    D 5 yards

19. Kendra is experimenting to find a design for an art project. She made these two figures on dot paper.

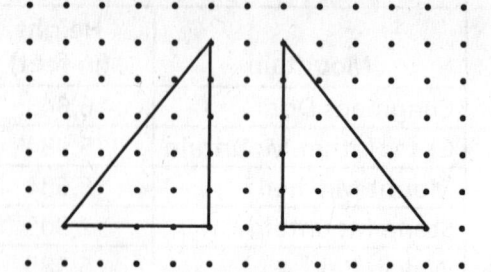

    Which transformation did Kendra use when she made the figures?

    A 180° rotation   C reflection
    B 90° rotation   D translation

20. A walk-a-thon is divided into 3 sections. The first section is 9.5 miles. The second is 11.7 miles, and the third section is 8.2 miles. Which is the *best* estimate for the total length of the walk-a-thon?

    A 12 miles   C 36 miles
    B 30 miles   D 52 miles

21. Joan and Marie shared a pizza. Joan ate $\frac{2}{8}$ of the pizza. Marie ate $\frac{1}{8}$ of the pizza. How much of the pizza was left?

    A $\frac{1}{4}$ C $\frac{5}{8}$
    B $\frac{3}{8}$ D $\frac{7}{8}$

Go to next page ▶

EOG PRACTICE TEST

Name _____

22. If twin Holstein calves together weigh about 755 pounds at birth, *about* how much does each calf weigh at birth?

    A between 100 and 200 pounds
    B between 200 and 300 pounds
    C between 300 and 400 pounds
    D between 400 and 500 pounds

23. Cammie and Corinne have the same amount of money. Cammie has 3 dimes. Corinne has 6 nickels. How much money would they each have if they doubled the value of their money?

    A 60¢
    B 36¢
    C 18¢
    D 9¢

24. A petting zoo has the shape shown below. How many feet of fencing is needed to enclose the shape?

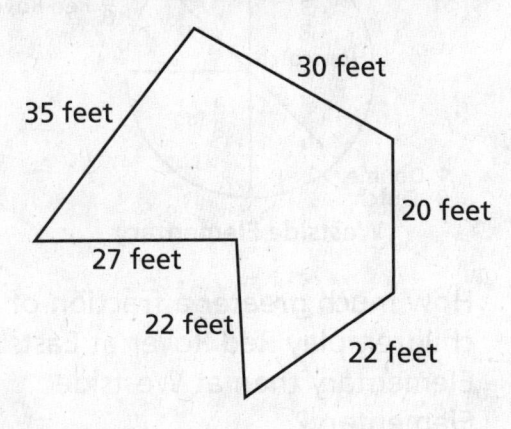

    (Shape with sides: 35 feet, 30 feet, 20 feet, 22 feet, 22 feet, 27 feet)

    A 156 feet
    B 116 feet
    C 110 feet
    D 105 feet

25. The distance from Los Angeles to Santa Fe is about 603 miles. If it takes a family 12 hours to drive the distance, *about* how many miles per hour are they averaging?

    A 35 miles per hour
    B 40 miles per hour
    C 45 miles per hour
    D 50 miles per hour

26. The Idle Hands Hobby Shop has 78 puzzles. It sells 29 puzzles each day for 2 days. Which expression shows how many puzzles are left?

    A $78 - (29 \times 2)$
    B $29 + 78 - 2$
    C $(78 - 29) \times 2$
    D $(78 - 2) + 29$

27. The table shows the outcomes for a game in which you use a spinner with 3 equal parts colored blue, green, and pink and toss a cube labeled 1, 2, or 3. How many possible outcomes are there?

    Find the Prize

Color	Number		
	1	2	3
Blue			
Green			
Pink			

    A 6
    B 9
    C 12
    D 15

Go to next page ▶

EOG Practice Test   Practice for the EOG Test

28. Which ordered pair identifies the location of the Weather Lab on the floor plan below?

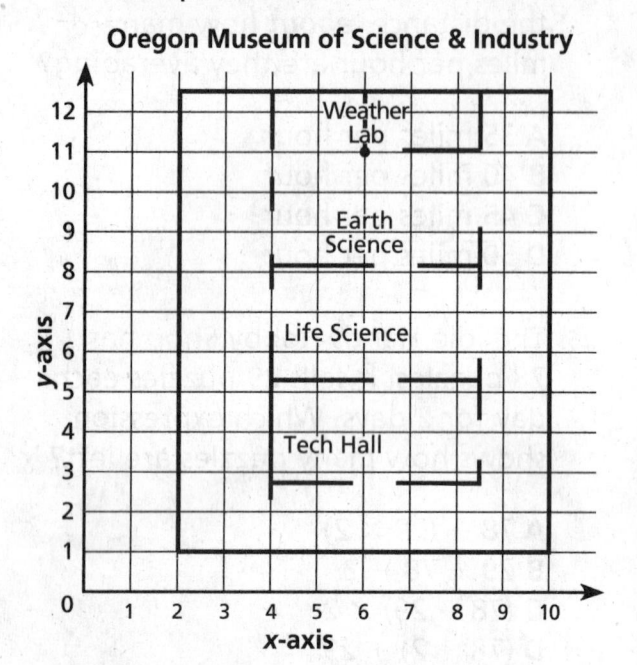

Oregon Museum of Science & Industry

A (11,6)   C (10,1)
B (5,11)   D (6,11)

29. Anne paid for a $5.65 lunch with a $20.00 bill. How much change did she get back?

A $25.65   C $14.35
B $15.00   D $12.35

30. The Blue Warbler Trail at Mineola State Park is 4.7 miles long. How long is the trail rounded to the nearest mile?

A 3 miles
B 4 miles
C 4.5 miles
D 5 miles

31. Maddie weighed 123 ounces at birth. Which weight is equivalent to 123 ounces?

A 6 pounds
B 6 pounds 11 ounces
C 7 pounds 11 ounces
D 8 pounds

32. The circle graphs show the makeup of different games children play at two different schools. The number of children at both schools is the same.

Games Played

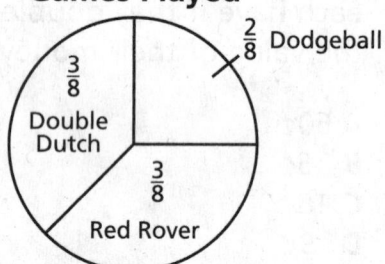

Eastside Elementary

Westside Elementary

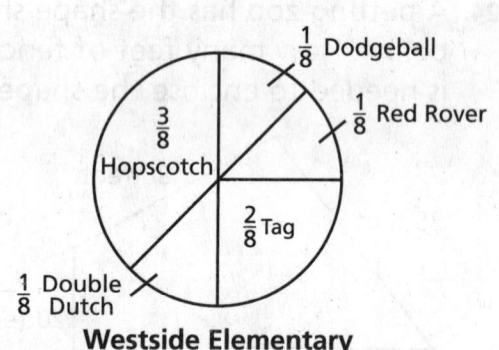

How much *greater* a fraction of children play Red Rover at Eastside Elementary than at Westside Elementary?

A $\frac{1}{2}$

B $\frac{3}{8}$

C $\frac{1}{4}$

D $\frac{1}{8}$

Go to next page ▶

33. The data in the table show the area of six states in square miles.

Land Area

State	Area (in square miles)
Michigan	56,803
New York	47,213
North Dakota	68,975
Ohio	40,948
Pennsylvania	44,816
Washington	66,544

Which state is about 12,000 square miles larger than Pennsylvania?

A New York
B Washington
C Michigan
D North Dakota

34. Zoe recorded the temperature outside her house each morning for five days. According to the graph, on which two days was it the same temperature at 6:00 A.M.?

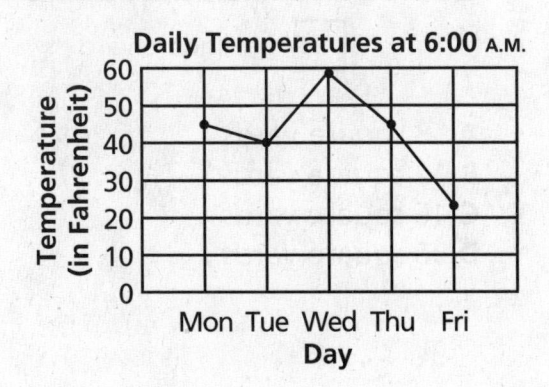

Daily Temperatures at 6:00 A.M.

A Monday and Thursday
B Monday and Wednesday
C Tuesday and Wednesday
D Thursday and Friday

35. The table shows how many loaves of bread a factory produces each minute. Let m represent the number of minutes and l represent the number of loaves of bread produced. Which is a possible rule for the pattern in the table?

Input	m	2	4	6	8
Output	l	20	40	60	80

A $m = l \times 10$
B $l = m \times 10$
C $l = m + 20$
D $m = l - 20$

36. Angie is biking on a trail. The trail is marked with signs like the one below. How many lines of symmetry does the sign have?

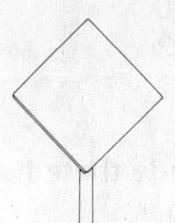

A 6 C 4
B 5 D 3

37. Melissa volunteered to pass out fliers about the school fair. She has 375 fliers to pass out. If she passes out 25 fliers an hour, how many hours will it take her to pass out all the fliers?

A 25 hours
B 20 hours
C 15 hours
D 10 hours

Go to next page

Name _____

EOG PRACTICE TEST

38. A new package of paper cups has 60 cups. *About* how many cups are in the opened package?

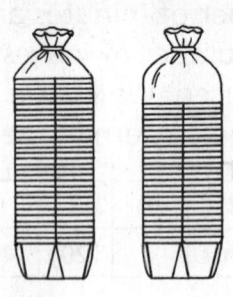

    A 50   C 30
    B 40   D 20

39. The total area of Hawaii is six thousand, four hundred twenty-two square miles. Which is the area of Hawaii written in standard form?

    A 60,422
    B 60,420
    C 6,422
    D 6,042

40. Dan made these two shapes on dot paper.

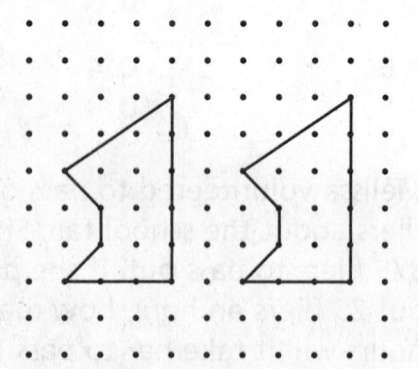

    Which transformation did Dan use when he made the two shapes?

    A translation
    B reflection
    C 90° rotation
    D 180° rotation

41. Alfonso wants to fill a wading pool with water for his little sister. Which is the *most* reasonable amount of water needed to fill the pool?

    A 25 cups
    B 25 pints
    C 25 quarts
    (D) 25 gallons

42. A gazebo in Watertown Square is in the shape of a hexagon. Use the diagram below to determine the *best* estimate of its area.

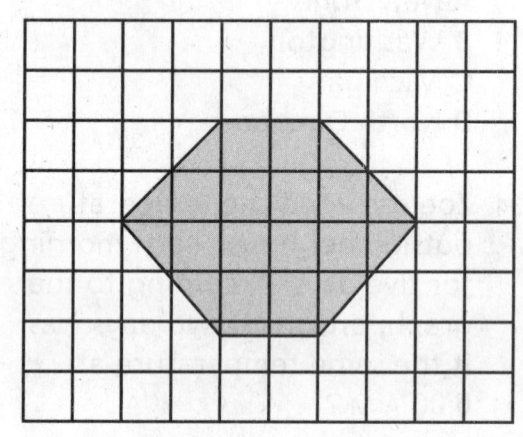

    ☐ = 1 square yard

    (A) 8 square yards
    B 12 square yards
    C 16 square yards
    D 30 square yards

    Go to next page ▶

124 Practice for the EOG Test   EOG Practice Test

43. Northbrook School is collecting soup can labels, which can be exchanged for school supplies. The students set a goal of 20,000 labels. If 400 students each collect the same number of labels, how many labels will each student collect?

   A 5
   B 50
   C 500
   D 5,000

44. Every time Sonia puts some money into her savings account, her parents add some money to it. Look at the table. Let s represent the amount of money Sonia put into her savings account and t represent the total amount of money. Which equation shows a possible rule Sonia's parents use to add to her savings?

Input	s	$10	$22	$28	$33
Output	t	$15	$27	$33	$38

   A $t + s = 5$
   B $t = s + 5$
   C $t = s - 5$
   D $s = t + 5$

45. In a carnival game, a player must reach into a top hat and pull out a red tag in order to win. If the hat contains 9 blue tags and 1 red tag, which is the probability of pulling out a red tag?

   A $\frac{50}{50}$
   B $\frac{9}{10}$
   C $\frac{1}{9}$
   D $\frac{1}{10}$

46. Northfork's baseball stadium is closed for remodeling. Before it closed, it held 42,500 people. To find the number of people it will hold when it reopens, change the digit in the thousands place to a digit that is 2 times the digit in the ten thousands place. How many people will it hold when it reopens?

   A 42,580
   B 42,800
   C 44,500
   D 48,500

47. Mark took a survey of his classmates to find out what colors of winter coats they owned. He used the data to make the double-bar graph below.

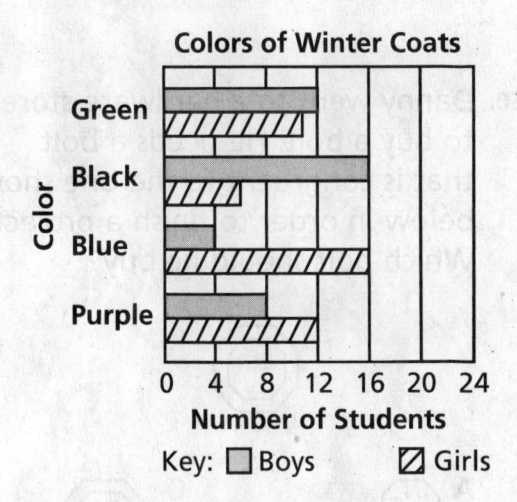

   Colors of Winter Coats

   (Bar graph showing Green, Black, Blue, Purple on y-axis; Number of Students 0-24 on x-axis; Key: Boys, Girls)

   Which two colors of coats are owned by fewer girls than boys?

   A green and blue
   B black and blue
   C green and black
   D purple and blue

Go to next page ▶

EOG Practice Test

Practice for the EOG Test 125

48. During his career, Al Spalding's average was 0.49 for the fewest walks per nine innings. Which fraction is equivalent to Al's average?

A $\frac{49}{1}$

B $\frac{49}{10}$

C $\frac{49}{100}$

D $\frac{49}{1,000}$

49. If 52 different people came to a puppet show every day for 30 days, how many people saw the show?

A 150
B 520
C 1,520
D 1,560

50. Danny went to a hardware store to buy a bolt. He needs a bolt that is congruent to the one shown below in order to finish a project. Which bolt should he buy?

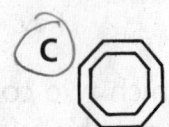

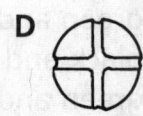

51. Mrs. Estefan drew a picture on the board to show how she lined up her flower pots along the edge of her porch. Which multiplication property did she illustrate?

A Zero Property
B Identity Property
C Associative Property
D Commutative Property

52. The Missouri Botanical Garden sells a book called *The Private Life of Plants* for $27. The garden received an order from a school for 28 copies of the book. What will the total cost of the books be?

A $756
B $560
C $540
D $415

53. The line plot below shows the number of television shows watched regularly by Ms. Levine's fourth-grade class. Which is the range of the data?

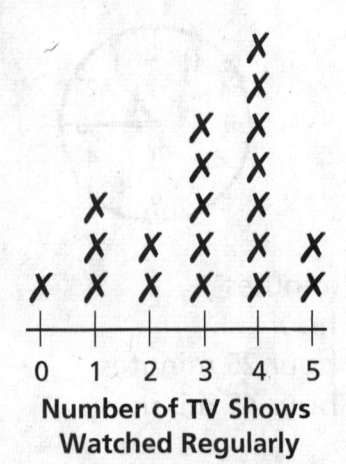

Number of TV Shows Watched Regularly

A 0
B 2
C 4
D 5 ✓

54. Anne has agreed to teach Mary how to ice-skate. She will charge $8 per lesson for 9 lessons. How much will Mary spend on ice-skating lessons?

A $81
B $72 ✓
C $64
D $56

55. Kishi could not decide how to arrange a leaf collection on a page in her scrapbook. She tried 5 rows of 6 leaves and then 6 rows of 5 leaves. Which multiplication property did she use to help her decide how to arrange the leaves?

A Zero Property
B Identity Property
C Commutative Property ✓
D Associative Property

56. Last January and February a total of 13.4 inches of snow fell. This year, 6.74 inches of snow fell in January and 4.82 inches of snow fell in February. *About* how many more inches of snow fell last January and February than this January and February?

A 3 inches ✓
B 2 inches
C 1 inch
D 0 inches

57. John carries a large ring of keys. To use one, he turns it clockwise and puts it into the keyhole, leaving it attached to the key ring. Which transformations is John using?

A reflection, translation
B rotation, translation
C rotation, reflection
D translation, translation

Go to next page ▶

EOG Practice Test

Practice for the EOG Test 127

Name _____

EOG PRACTICE TEST

58. Mrs. Koch has $208 set aside to give to charities. She wants to donate $8 to each charity, so she divided $208 by $8. She says she will be able to donate to 26 different charities. Which operation can she use to check her statement?

  A division
  B multiplication
  C addition
  D subtraction

59. A class of 20 voted for which camp they wanted to go to for an outdoor field trip. $\frac{2}{10}$ chose White Pines. $\frac{3}{5}$ chose Spruce Hills, and $\frac{4}{20}$ chose Lookout Point. No one chose Forest Knoll. Which two camps got the same number of votes?

  A White Pines and Spruce Hills
  B White Pines and Lookout Point
  C Spruce Hills and Lookout Point
  D Spruce Hills and Forest Knoll

60. Andy drew this design. What transformation did he use to make the pattern unit?

  A reflection
  B translation
  C 90° rotation
  D 270° rotation

61. Ethan's sister got on a train at 1:15 P.M. to take her home from college. She arrived home at 2:40 P.M. How long did her trip take?

  A 25 minutes
  B 1 hour
  C 1 hour 25 minutes
  D 1 hour 45 minutes

62. Ella's dad is standing next to a statue. He is 6 feet tall. **About** how tall is the statue, including the base?

  A 9 feet
  B 12 feet
  C 18 feet
  D 50 feet

Go to next page ▶

128 Practice for the EOG Test

63. Which number makes this number sentence true?

$$(4 \times 2) \times 5 = 4 \times (2 \times \square)$$

A 2
B 4
C 5
D 10

64. A computer game shows this figure making this move. What transformation does the moving figure make?

A 90° rotation
B 180° rotation
C reflection
D translation

65. There are 4 people making 6 finger puppets each. There are a total of 24 finger puppets. Deanna wrote the equation below to match this statement.

$$4 \times 6 = 24$$

Which of the following is a related equation?

A $24 \div 6 = 4$
B $24 \div 2 = 12$
C $24 \div 12 = 2$
D $24 \div 3 = 8$

66. The Best Friend's Pet Hotel serves 3,268 pounds of dog food over an 8-week period. During the same period, it serves 2,785 pounds of cat food. How much pet food does the pet hotel serve during an 8-week period?

A 5,943 pounds
B 5,953 pounds
C 6,043 pounds
D 6,053 pounds

67. Dominic bought 8 notebooks. He paid $16 for the notebooks. Let c represent how much each notebook cost. Which is the solution for this equation?

$$16 \div c = 8$$

A $c = 1$
B $c = 2$
C $c = 4$
D $c = 8$

Go to next page ▶

EOG Practice Test   Practice for the EOG Test   129

Name _____

EOG PRACTICE TEST

68. Kim moved the puzzle piece from position A to position B. Which transformation did she use to move the puzzle piece?

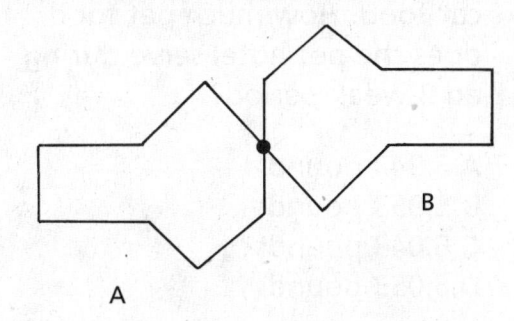

   A 90° counterclockwise rotation
   B 180° counterclockwise rotation
   (C) reflection
   D translation

69. Jessica made this model to show 0.80. Which of the following is equivalent to 0.80?

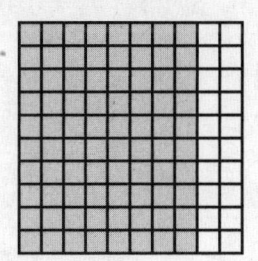

   A 0.08
   B 0.8
   C 8
   (D) 80

70. The Geo Rock Museum has a collection of 856 rocks. If it displays an equal number of rocks in each of 25 display cases, how many rocks will not be displayed?

   A 0
   (B) 6
   C 12
   D 24

71. Rosa walked $\frac{4}{10}$ mile from her home to her friend's house. Then she walked $\frac{3}{10}$ mile from her friend's house to a store. Which model shows how far Rosa walked altogether?

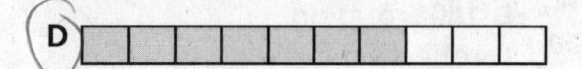

   A
   B
   C
   (D)

72. Christine wants to buy 8 large photo books of birds for gifts. Each book costs $75. Which equation could she use to help her find how much the books will cost altogether, using mental computation?

   A $(70 \times 8) - (5 \times 8) = 520$
   B $(70 \times 8) + (5 \times 8) = 600$
   C $(70 \times 4) + (5 \times 4) = 300$
   D $70 + 5 \times 8 = 115$

Go to next page ▶

130 Practice for the EOG Test

EOG Practice Test

Name _____

EOG PRACTICE TEST

73. The table shows the diameters of some of the planets in our solar system.

Diameter of Planets

Planet	Diameter (in miles)
Earth	7,926
Jupiter	88,734
Neptune	30,199
Saturn	74,566

Which three planets are the largest and arranged in order from *largest* to *smallest*?

A Neptune, Saturn, Jupiter
B Saturn, Neptune, Jupiter
C Jupiter, Saturn, Earth
(D) Jupiter, Saturn, Neptune

74. Emilio planted 12 rows of tulips in his flower garden. He planted a total of 132 tulips. How many tulips did Emilio plant in each row?

A 8
B 9
(C) 11
D 12

75. Jean and Julia each brought $3\frac{3}{4}$ cups of snack mix to a slumber party. How much snack mix did they bring in all?

A $7\frac{1}{2}$ cups
(B) $6\frac{3}{4}$ cups
C $6\frac{1}{2}$ cups
D $4\frac{1}{2}$ cups

76. Coach Rogers timed his top four runners in the 100-meter dash. He recorded their times in the table below.

100-Meter Dash

Runner	Time (in seconds)
Jackson	15.07
Miller	15.71
Tutland	14.85
Bozak	15.79

Which set of times shows how the runners placed from the fastest time to the slowest time?

(A) 14.85, 15.07, 15.71, 15.79
B 15.07, 15.71, 15.79, 14.85
C 15.79, 15.71, 15.07, 14.85
D 14.85, 15.79, 15.07, 15.71

Go to next page ▶

77. Shawna has 38 shells. She wants to know how many will be left over, if she puts 11 shells into each of 3 boxes. Which expression can Shawna use to find out?

A (38 − 3) × 11
B 38 − 3 × 11
C 38 + 11 − 3
D 38 × 3 − 11

78. Hector's dad is helping him put up a tent. Which term identifies the relationship between the center pole and the ground?

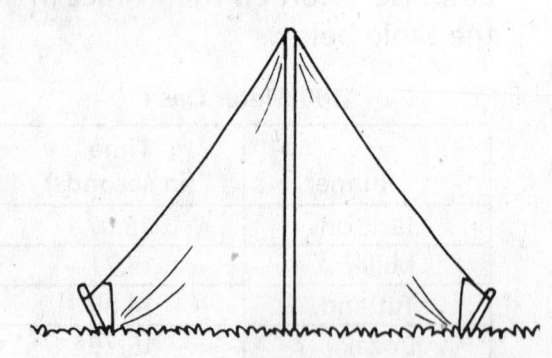

A parallel
B symmetrical
C congruent
D perpendicular

79. A catalog business has an order for 32 baskets with 19 bars of soap in each basket. How many bars of soap will be needed to fill the order?

A 596
B 600
C 608
D 640

80. One morning at Sam's Snacks, 58 people each ordered an 8-ounce glass of orange juice. *About* how many ounces of orange juice were ordered that morning?

A 100 ounces
B 400 ounces
C 480 ounces
D 700 ounces